The ... **hey saw...**

...ky was now filled with *hundreds* of shiny objects flashing signals to one another!

Together they sat holding each other, too frightened to move. Too frightened until it began. Slowly at first. Like parachutes floating, free-falling downward so gracefully, so easily that it seemed indiscernible at first. So they stared; captivated by the prospects as dozens, no, hundreds of the glowing white objects travelled like falling stars down and across the black, onyx sky. Subtle, graceful, beautiful in their way, the round and shining orbs were descending upon them...

"Highly unusual, traumatizing events outside the realm of normal human experiences."

—Bernard Vittone, M.D.,
National Center for Psychiatric Disorders,
Washington, D.C.

"The most compelling case of its kind
I have ever encountered."

—William L. Anixter, M.D.,
Mountain Psychiatric Center,
Asheville, North Carolina

SEARCHERS

A TRUE STORY

RON FELBER

ST. MARTIN'S PAPERBACKS

This book for Laurie,
Chris and Gregory;
perfect.

The names of several individuals mentioned in this book have been changed to protect their privacy.

SEARCHERS

Copyright © 1994 by Armstrong Publishing Group Ron Felber.

ISBN: 0-312-95511-1

Printed in the United States of America

Armstrong Publishing Group hardcover edition published 1994
St. Martin's Paperbacks edition/July 1995

St. Martin's Paperbacks are published by St. Martin's Press, 175 Fifth Avenue, New York, NY 10010.

10 9 8 7 6 5 4 3 2 1

PREFACE

What follows is an incredible true story. If these adjectives seem a contradiction, they are not for the story I am about to relate defies both our senses and the reality we share, yet it is absolutely true.

In November, 1990, while I was on a business trip to California, Paul Moran, a longtime associate, told me about Steve and Dawn Hess and the trauma they endured while in the Mojave Desert some thirteen months earlier. It was over dinner and in confidence because few people knew of what had occurred.

Frankly, the Hesses were afraid to tell anyone outside their immediate family and Paul, Steve's closest friend, for fear of ridicule and more...Nevertheless, Paul felt it important that I know. As a writer, the story would be of interest to me and Paul knew it. As a friend of the Hesses, he hoped that having someone listen to and communicate their astounding encounter would aid them in at last coming to grips with the experience.

Steve and Dawn Hess are stable, credible individuals. Steve is a thirty-two year old supervisor of large construction projects. He is a graduate of the University of Redlands where he excelled in football as an all-league linebacker in the early 1980's. One's immediate impression upon meeting him holds. Steve is a quietly confident, self-reliant man. The kind of witness any attorney would relish having on his side in a case where the outcome hinged on a single individual's testimony.

As a person, Dawn Hess is no less impressive. Like her

husband, she is the product of a middle class upbringing and a University of Redlands graduate. Dawn is the mother of three; Steven, age 6; Bethany, age 4; and Amberly, 22 months. She is gentle, yet strongwilled, particularly when it comes to her family. The more sensitive of the two, Dawn is the recorder of fine detail; the problem solver confronted on the evening of October 22, 1989, with a mystery she and Steve will be forced to contemplate for the rest of their lives.

This is their story. A crack in the wall, if you will, that offers the rare opportunity to see beyond this reality into another so startlingly different that it will change the way you view man's place in the world forever.

R.F.
Chatham, New Jersey
June, 1993

"The moon shall bleed
its light,
And the stars shall fall
from the heavens."

Revelation 6:12-13

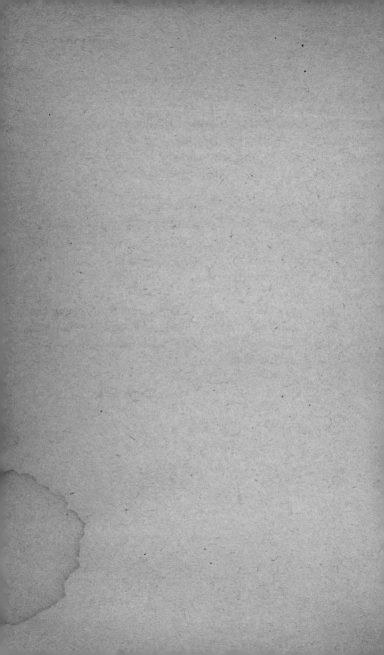

I: INVASION

When we first saw them dropping from the sky we thought it was some kind of military manuever; maybe for Operation Desert Storm. But it was too massive even for that. I mean, there were thousands of them falling, then rushing toward us.

So I kicked out the campfire, grabbed my gun, and ran into the back of the camper with Dawn. Then we sat there, Indian-style, waiting. Until they came. Thousands of them. Thousands of pairs of tiny, red eyes glowing in the dark around us.

Steve Hess

ONE

La Mirada, California
October 20, 1989
12:20 p.m.

The getaway weekend Steve had promised seemed godsent, Dawn was thinking as she vacuumed the wall to wall carpeting in their living room. Outside, Steve was packing the 1987 Ford pick up his Dad had lent them for the trip. Complete with campers' shell, a double bed and fully carpeted interior, she was hopeful its conveniences would make their weekend in the Mojave Desert more bearable.

Originally, it had started as a hunting trip, an idea she found abhorent, but now their plans had expanded to include a stop at Cima Dome and sightseeing at Mitchell Caverns. Not exactly a jaunt to the Islands, but these days with the kids and Steve's hectic schedule, just getting away together was enough to satisfy her. Besides, they'd be camping at Midhills which Steve knew to be clean and well cared for, so it wouldn't really be roughing it.

She flicked off the Electrolux. Her eyes scanned the living room for toys, baby bottles and Zweiback toast; the artifacts a young mom came to expect after having scoured their four bedroom ranch-style home for the better part of the morning. Call her the nervous type or just plain conscientious, but she wanted the house to look right. Afterall, Steve's parents weren't obliged to babysit and agreeing to take the kids for the weekend was a big favor. The least she could do was to see to it that the place was clean, with beds made and dishes done. She bent down to retreive a 'transformer' little Steve had tucked away beneath the coffee table.

Dawn turned the plastic toy inward so that the spaceship converted into a tiny, green monster howling up from the center of her palm. The face was twisted grotesquely, whether in fierceness or in pain she could not tell. She studied it for a drawn moment. *Why on earth would a silly toy like this seem so strangely significant?* she wondered before tossing it into the toy chest in the corner of the room.

"Steven! Steven Ray, where are you?" she called. Dawn Hess padded through the hallway and into the baby's bedroom looking immediately to the crib where Bethany, their six week old, lay fitfully sleeping. Her hair was blonde and her skin fair like Dawn's. Her legs and body were long and extended beneath the cotton blanket. 'Gonna make a fine sprinter someday,' Dawn reflected in a flash of memories back to her track days at Chaffey High. 'A natural born runner.'

The thought had no sooner passed through her mind when she caught sight of Steve, Jr. Already a climber at 2 1/2 years of age, he was on the other side of the crib attempting to scale the guardrail! She rushed across the room in time to catch him, then shook her head at the futility of it all.

"You little monkey!" she scolded holding his face six inches from her own. "You were gonna pounce on your baby sister, weren't you?"

She rubbed her nose up against his. He giggled and squirmed as Dawn held him back from her again, staring deep into his crystal blue eyes. For a split second she thought she saw something in them; no *behind* them, that caused her pause. Quite unlike his own, these appeared the wizened, knowing eyes not of an infant, but of an old man. 'We know one another well,' the eyes seemed to be saying. 'We are the same, you and me, in mind and heart and soul.'

These reflections were interrupted by the sound of Steve's heavy footfall as his 6' 1", 225 pound frame ambled down the hallway toward Bethy's room.

"Where the hell are they?" he pled in exasperation.

"Shhh," Dawn hissed back at him. "The baby!"
He held up his right hand in a gesture of compliance. "Sorry, I didn't know she was sleeping..."

She screwed her eyes to the ceiling, feeling every minute of their five-year marriage, then gathered Steve, Jr. in her arms.

"Have you called?" she asked once they'd entered the hallway.

"Yeah, yeah, I called. There's no answer."

She planted little Steve feet-first on the carpet, then bent down to tie his shoe.

"Then, they're on their way. There's no need to have a heart attack about it! You'll get to shoot your deer or buck or whatever it is you're after. I swear, sometimes I think you're more of a little boy than your son!"

"It's a buck," he pouted. "The kind of four-point buck I've been after every year for fifteen years out in the Mojave!"

Steve marched from the hallway into the foyer, then out the front door. Steve, Jr. dawdled after him as Dawn watched, still on one knee. Men she swore quietly to herself.

Once outside, Steve walked to the back of the truck in the driveway. He extended the tailgate reaching beyond the gas stove, blankets and cooler to grab the 7mm Browning Magnum that hung on a rack above the folddown bed on the camper's right side. He held the weapon up, sight to eye, then took aim. In his mind's eye he could see the mule deer, near big as a horse, like he'd seen it one week before. Big as it was, tracking the huge animal down and locating it once you did was no easy feat. Its grayish color was indistinguishable from the pinyon pine, shrub pines and high sand; so much so that more than once he'd heard of hunters suddenly discovering a three hundred pound buck not fifteen feet from them!

But it wasn't like that when Ron, his 26 year old brother, Dad and he were out in California's high desert near Tabletop Mountain at the opening of hunting season. No, he had the buck all right. Not one hundred yards off in the distance, poised like a statue on a knoll just north of them. Since he'd found it, it was his shot. His chance to bag a magnificent specimen, huge with a four-point rack of antlers, directly in his gun sight. But he missed! Goddamn failed, he cursed, letting the weapon fall back down to his side; and he wasn't used to failing; not in football where he played first team middle linebacker for the University of Redlands; and not in business where, as Project Manager for Southwest Construction, he never missed an opportunity for a sale, always bringing his project in on time and under budget! Hunting was no different. It was determination that paid off in the long run, Steve always liked to say, and he was out to prove it again this weekend.

Steve returned the Browning to its rack beneath the shell of the camper. Little Steve had climbed onto the tailgate and was worming his way through the menagerie of camping equipment. Steve took him into his arms.

"Your Daddy's gonna bag that deer, Stevie," he promised. "You wait and see. On Monday night, after we come back, the whole family's gonna have one hell of a venison dinner."

It was then that Steve heard the long-awaited sound of his Dad's Chevy Lumina parking curbside. The engine stopped, two car doors opened and slammed as Steve, Jr. wrestled his way out of Dad's arms and down to the ground.

"Jeez, what took you so long?" he called.

His Dad, whom they called Wolfy because of his white hair and sharp, wolfish features, smiled broadly, proud to display the overspill of paunch that had crept over his beltline during the past couple of years. He collected little Steve into his arms.

"Barstow's not around the corner, you know. We hit traffic on I-15. What's the hurry anyway?"

"Easy for you to say," his son joked. "All you've got to look forward to is babysittin'. I'm on my way to bag a four-pointer!"

"I wish you luck after last week. You deserve it, son."

Diane, Steve's Mom, came up beside Wolfy. He passed little Steve into her open arms. Tall and thin in contrast to her ex-Marine husband's stout frame, her red hair was cut short in a pixie style that flattered her longish face, high cheekbones and distinctive features.

"Well, if you do get that buck make sure he's gutted and cleaned before you bring him home. My God, the mess your father's put me through over the years with his hunting!"

Dawn appeared in the doorway opening the screen door halfway as Diane neared the entrance. "Hi Folks! Bethy's asleep, her formula's in the fridge and Steve's already eaten lunch."

The father and son watched as they disappeared into the foyer.

"You know, Dad, if it's all the same to you, I'd just as

soon get on the road. We're all packed and it's already past noon."

Wolfy just nodded knowing how his son could be, once he'd gotten an idea in his head.

Inside the house, Dawn leaned over the safety rail of Bethany's crib. She kissed the sleeping infant tenderly as Diane looked on from the doorway, Steve, Jr. at her side.

"Come on, Dawn, it's showtime!" her husband's baritone boomed from the foyer.

She straightened up, then gave her mother-in-law a look of exasperation. Walking from the room, Dawn took Steve, Jr. in her arms.

"Now you be a good little boy for your Grandma and Wolfy," she instructed.

"Mommy go?" he asked.

"Yes," she answered, the uneasiness of moments before returning, "Mommy go bye-bye."

12:45 p.m.

Steve and Dawn were all too familiar with the snarls of stop and go traffic on the 605 Freeway and like most southern Californians had learned to take it in stride. In the forefront of Steve's mind was the freedom this long weekend had come to represent. Shuttling between jobs at the Norwalk State Hospital and ongoing Metrorail construction in Long Beach had left him tapped out emotionally. Traffic, budgets, unions — Jesus, the list of problems seemed endless. Yet, here he was about to break loose from the tentacles of L.A.; way loose to the familiarity and calm of the Mojave. And underlying it all, he thought, a smile of anticipation passing over his lips, was the buck. Totally unlike the day-to-day bureaucracy he'd become embroiled in with the state and local governments at work, here was his chance to hit the mark; to crack that sucker and take him out once and for all.

He glanced momentarily to Dawn who was locked in her own reveries of kids and how she and Steve had finally decided on the trip. It was the Thursday before, she recalled, when his parents, who lived in Barstow, offered to babysit for them. They'd gone to dinner at Rosita's, a neighborhood restaurant, when after two Margueritas, Steve broke the news.

The week before he'd missed the opportunity of a lifetime, he'd told her, and wanted desperately to go after that buck. What's more, he was leaving this weekend and wanted her to come along!

Dawn's first response was 'no' on principle because she genuinely disliked the idea of tracking some poor animal through the desert, then killing it. But Steve could be persuasive and felt guilty about leaving her home for weekends back-to-back. "Point is, it doesn't have to be just hunting," he'd promised. "There's Cima Dome, Tabletop Mountain, Mitchell Caverns. I tell you, Dawn, staring down from on top of those mountains is like standin' on the edge of the world."

No doubt he was right. There wasn't anyone she knew who understood more about the Mojave than Steve. So, given the fact that his parents had already agreed to take the kids and that she could use a break from the house anyway, she agreed.

Dawn gazed from out of the passenger window. They were headed east on I-15. Already the roadside was taking on the tan, barren look of a desert.

"Nice of my folks to watch the kids like this, eh babe?"

"Awful nice," said Dawn stretching her long arms to

the trucktop and yawning. "I just hope they behave; especially Steven."

"He's at that age all right. Into everything."

"Yep. But you know lately, it's more than that; something different I've been noticing. I'm not sure what." Dawn reached behind the front seat and into the cooler for a can of Coke. She pulled open the fliptop, then took a sip.

"You know how close we are; especially since he was sick with that heart problem."

"Yeah..."

"Well, you may not believe this, but lately I can tell what he's thinking. Honest. And know what else? He can tell what I'm thinking, too. Even before we speak. He looks at me and I know that he understands. We just do it, or get it, with no words exchanged. We just know."

Steve took a gulp of soda, then depressed the gas pedal down to the floor as traffic became non-existent.

"I suppose that most moms feel that way about their kids."

"No, not like this, Steve. It's deeper than that; it's in his eyes and just not the same at all with Bethy."

"Hey, I believe you!" Steve clowned, seeing the seriousness in her eyes. "Honest! I'm just a guy tryin' to shoot a deer this weekend. Overspent budgets, town council meetings, clairvoyant kids...all of it fifty miles away and likely to get a lot farther." He bent down to produce a glass jar filled with quarters from beneath the truck seat. "But in case you forgot this is a vacation, I thought we might stop at *Whiskey Pete's* to try our luck at the slots before we turn in for the night at Midhills." The gravity of Dawn's concerns seemed to lift like fog in the midday sun. She leaned over to give him a peck on the cheek.

"Sounds fine to me," she revelled, taking a swig of

Coke. "Let's take these quarters and go win us some money!!"

The driving time that followed was a period of introspection for both of them. Certainly there were times during their five years together when Dawn and Steve wondered if their marriage would last, and might not have, had it not been for Dawn's Mormon upbringing and devotion to family. Early along there were the finances that left them barely able to make their rent each month. Then came baby Steven and his hospitalization, the loss of Dawn's income and most recently Steve's long hours on the road and at work. These moments of doubt were few and, in the long run, left them closer than ever; but there were other occasions of equal intensity when both knew they were meant to spend their lives together. The birth of Bethany six weeks earlier and the last half year generally was a time like that. Fact is, at this early stage of life, they had it made. Steve's work was demanding, but he'd have it no other way, little Steve's health was back 100% and their marriage seemed to have broken through the early hurdles to be tempered, strong and enduring.

Such was their disposition as they grabbed a late dinner at a steakhouse near Nipton, then crossed the border into Nevada. This was their time to let loose and they did, Steve swigging a Coors as he played blackjack and Dawn sipping from a Long Island Iced Tea as she tried her hand at the slots. In the lounge, a country western band blasted Charlie Daniels rockabilly, which was just fine with Dawn, who loved to dance and found Steve a willing partner once he'd won $150 at the tables.

Listen to music, gamble and dance to the early morning; not a bad way for two homebodies to begin three days away from the ratrace existence of southern California. By

2 a.m., exhausted and $70 richer, Steve and Dawn began their double back from Nevada to the Midhills camp site some ninety miles away.

"Well, we won, didn't we?" Dawn asked, the excitement of the night still with her.

"Yeah, we did," Steve answered absently, trying to navigate the fog and rain that shrouded the I-15.

"Weather's sure changed."

"Like that in the desert. Especially this time of year."

Dawn shivered. She folded her arms close to keep warm, peering idly from out of the passenger side window as they turned off the highway headed toward the New York mountains. Over the past four hours the temperature had dropped nearly thirty degrees and a driving wind had pick up, pelting the truck with sand as it knifed through the thick, clinging mist.

An eerie chill passed through her as she gazed beyond the rafts of fog to the silhoutte of volcanic formations. They rose up from the desert floor like enormous waves of earth, accentuated by an occasional Joshua tree whose spiked branches jutted skyward like clawing fingers into the night. Steve had told her about the mining operations that flourished here in the 1800's and how even now this alien terrain of canyons and craters was the source of cerium, lanthanum and dozens of other rare earths with science fiction names related to high tech super conductivity.

Her thoughts were shattered by the screeching sound of burning rubber as she was thrown forward into the dashboard.

"Jesus, Steve!"

The truck stopped dead.

"Sorry. We must have hit a washed-out section of the

road. With the fog it's tough to see."

"How much farther to Midhills?"

"Too far to go in this weather." He hesitated. "I was thinkin' it might be best to pull over and get some sleep. By morning the sun will have burned all of this fog off and driving will be alot easier."

The high she had been on earlier had evaporated leaving Dawn tired and groggy.

"I suppose. If that's what you think's best..."

Steve pulled the truck off the canyon road not fifty feet from the foothills. He exited, then made his way around as Dawn climbed into the back of the camper. Lights on, she cleared the bed while Steve surveyed the surrounding area. Tucked as they were off the road and nestled between two granite and limestone mountains he turned, satisfied they were safe, then re-entered the truck from the back.

Dawn had already washed with bottled water and was sitting on the edge of the bed when Steve's bearish frame filled the back of the camper.

"I don't know what you have in mind, but I'm sleeping," she said as she he began undressing.

"Fine by me," he muttered, setting an Ithaca 12 gauge shotgun beside him. "I'm too tired to even think about it..." He pondered a moment. "But, tomorrow..."

She didn't let him finish.

"Tomorrow is another day."

Steve got under the sheets. Dawn, who was of a different sort, could not retire so easily. Like a cat, she would explain to Steve, she needed to get the feel of a place. And so she did, still battling a sense of foreboding that it seemed would not quit.

She peeked from out of the camper's long horizontal side window beyond the rugged desert brush along the foothills.   Her eyes travelled straight up the jutting mountainside to the volcanic peaks that shown like castle spires above the dense fog. She shook her head at the unreality of it all and was about to comment on it to Steve when she realized he was asleep. She smiled warmly at the sight of her exhausted husband snoring, gun in the ready, her own sleeping sentry.

At last, Dawn pulled the curtains closed. She flicked off the interior light and was about to crawl beneath the covers when she felt a chilling sensation quite unlike any she had ever experienced. She surveyed the camper from gas stove to backpacks, to sleeping bags when her eyes fell upon it: The transformer.

Inverted, the howling creature stood staring up at her; cold and stark as a severed appendage.

TWO

New York Mountains
October 21, 1989
7:30 a.m.

Pencil thin rays of sunlight groped through the clouds still remaining from the night before. Dawn yawned and stretched lazily. From out of the back window she could see Steve was already up and around kneeling over the gas stove; a pot of coffee brewing. He looked up and smiled. She waved, then opened the tailgate from within.

"You looked so content, I decided not to wake you."

"What time it it?"

"Seven-thirty or so. Want some coffee?"

She nodded, slipping off her nightgown and stepping into a fresh pair of panties. Steve poured a cup of hot coffee admiring the beauty of his wife. Her chestnut brown, shoulder-length hair, unbrushed and dishevelled, seemed sexy. Her body was toned and trim; her lithe figure nearly returned to the pre-baby days. Still athletic, he thought, a natural beauty. A fact that was driven home to him at moments like these when she was freshly awakened and without make up.

He walked toward the open tailgate, coffee in hand as she threw on a pair of jeans and a "U of Redlands" sweatshirt.

"Have I told you lately how pretty you are?" he asked.

Dawn took the coffee from him.

"Not often enough," she answered coyly, taking a sip.

He sat on the edge of the tailgate, feeling a twinge of inadequacy at how busy his work had kept him and how physically out of shape he had become by comparison.

"I know. The job has taken too much of my time. I haven't paid as much attention as I should to either you or the kids."

Dawn took him into her arms. They embraced.

"You're a good husband, Steve. There's nothing I'd change about you."

They kissed. She eased backward, then shimmied farther into the truck; a flood of emotions rising up within her. Steve, the athlete. Steve, the stoical provider who tried so hard and asked so little of everyone. This was a good man, her heart was telling her. In a quiet, unsensational way, he was everything she had ever wanted.

Steve lay down beside her. "I love you," he whispered.

They kissed deeply, then made love in the back of their Ford camper parked off an isolated canyon road miles from anyone in the high desert of the East Mojave.

Dawn brushed her reddish-brown hair looking in a mirror affixed to a curtain rod in the truck's interior as Steve began to pack up. He pulled a pair of binoculars from his backpack, then grabbed his 7mm Browning rifle; all essentials to the "off road" hunting they were about to begin. He made his way to the cab, hunting gear in hand while Dawn struggled with a tangle of hair. She studied herself in the mirror finally noticing the transformer she had discovered the night before laying beside their bed. Oddly, for all the impact it had had on her then, it seemed forgotten now and only vaguely real as if it had been some minor part of an elusive dream.

She put the hairbrush down, then took the tiny toy into her hand as Steve came around again to collect the canteen.

"Where did this come from?" she asked vaguely, holding the transformer up for him to see.

"Little Steve brought it to me yesterday just before we left."

"He did?"

Steve, who was paying more attention to his gear than the conversation, looked up.

"Yeah. What's the news in that? He's always handing me his toys so we can play together. You know that."

Dawn shrugged, converting the space creature to a spaceship and back again as she spoke.

"I don't know. It's just that I put it in the toychest before we left."

"So, he went in and got it while you were with Bethy."

"Did he say anything? When he gave it to you, I mean?"

Steve looked at her with a mixture of surprise and annoyance.

"He said, 'space man'. Just the kind of thing you'd expect a two year old to say when he hands you a transformer. So what's the big deal?"

"I, I don't know," she stammered, putting the anxiety she felt back in the box of her subconscious. "It's just these mountains, I guess. The desert. So far from everything. So strange looking. Kind of gives me the creeps."

Steve grinned, then kissed her on the forehead.

"Hey! There's no need to be frightened. You're with the 'desert rat'. The only thing you've got to worry about is what we're going to do with all the venison once I bag that deer I've been telling you about."

Dawn agreed. The brave trooper. She kissed him back, then left the camper, Steve watching as she walked beyond the foothills to the edge of the precipice where she flung the transformer as far as she could. Strangely gratified, she watched as the tiny creature dropped thousands of feet down to the desert valley below. It was gone, she thought then, gone and out of their lives forever.

Within the hour, the camper was making its way along

the rugged dirt road between the New York and Providence mountains. The weather was clear and sunny now, a cool seventy degrees with the air fresh and frothy, cleansed after last night's rain. Steve took the truck over backroads toward the Midhills camping grounds "off roading"; hunting off the main roads as Dawn, the scout, peered through binoculars searching for buck and bighorned sheep.

"Is that one?" asked Dawn pointing out some seventy-five yards north of them.

Steve reached for the binoculars easing his foot from the accelerator. He scrutinized the area as she continued to point beyond the gray-brown desert scrub to a cluster of yucca plants and the shadow of a juniper tree that fell upon it.

"That's no deer. It's a juniper tree. Shadows can fool you."

She took the binoculars back, then looked as if to confirm his judgment.

"Guess you're right," she concluded putting the binoculars down on the truck seat. "It all seems to blend, doesn't it?"

"At first. But the desert has as much life in it as anywhere else. Just more subtle with the different life forms all working together to survive." He pointed to another clump of trees and plantlife. "Take that Joshua tree. That tree gives shelter to desert wood rats, squirrels, and the like. Birds like shrike and pinyon jays make their nests in its branches. Lizards and desert tortoise get protection from the sun in the shade of its base. Even the yuccas that surround it kick in because yucca moth fertilize the tree's flowers whose seeds later become its food. All of it subtle; but there if you know what to look for."

"So, why do you want to alter that ecology by killing helpless desert deer?"

Steve looked to her bracing himself. It was a discussion they'd had before.

"I've been hunting with my Dad since I can remember, Dawn. Hunt or be hunted. That's the way it is. On the other hand, maybe we're just as much a part of the ecology as any other animal. Maybe hunters are part of what keeps it in balance."

"And who hunts us?" she challenged.

He reached toward his 7mm Browning wrapping his fingers around the midpoint of its barrel.

"No one," he answered, giving her his best smart-assed grin. "'cause we've got the brains and the weapons to get them first."

The two continued backroading south through the high desert woodlands honeycombed with abandoned mines and ghost towns to the basin-and-range terrain with its flattop peaks and bone dry riverbeds leading to the Providence mountains. Occasionally, Steve would stop the truck to hunt on foot leaving Dawn to relax in the back of the camper with a Coke or cup of coffee. By early afternoon with the temperature having risen into the eighties, they found themselves near a point of Steve's choosing beyond Midhills and just north of Mitchell Caverns.

"It's getting hot. Deer tend to be active dawn to noon, then settle down until it cools off around three or so." Steve pulled out a desert access guide spreading it across the front seat. "As I see it, we're here," he offered pointing to an area just off Black Canyon Road. "If we head up here toward Forshay Pass, we can have lunch at a place I think you might like."

"What place?"

He raised his palms in the air, staving off the inquiry. "It's a surprise."

With that, they made their way back onto the main road departing as quickly for a dirt path as rugged and overgrown as any they'd travelled earlier. Up the Providence mountains they drove beyond Forshay Pass to an elevation of nearly 7,000 feet where at last they stopped. Steve left the camper carrying a blanket and cooler filled with soda and sandwiches; Dawn followed, binoculars hanging from the leather strap around her neck; both ready for some sunshine and a leisurely picnic.

Steve walked up ahead of her the final twenty yards stopping at the brink of what must have been a drop of five thousand feet or more. He had put the cooler down and was spreading the blanket when Dawn arrived.

"So, what do you think?"

"Unbelievable!" she marvelled, her Reeboks making a semi-circle in the sand as she arced around to view the panorama of volcanic formations and desert floor that lay before them. "It's like the entire Mojave is right here at our feet."

"I know," he answered excitedly. "Out that way rising up from the ground like an inverted saucer is a seventy-five mile volcanic formation called Cima Dome; the most perfect natural sphere in the world." He sidled up next to her, pointing. "And just west of that is Cinder Cones. You see those peaks shooting out of the ground like that. They're actually extinct volcanoes, some of them formed less than a thousand years ago."

She craned her head around to him, still cradled in his arms.

"What else?" Dawn asked impressed. "What else do you know?"

"Well, there's more. Like how the East Mojave was formed 150 million years ago, the result of fantastic volcanic eruptions. You'd never know it now, dry as it is, but not so long ago this entire area was covered with rivers and lakes. There's still all kinds of prehistoric fossils around. But then, the weather became hotter and drier and just about everything that was alive shrivelled away and died."

"Darwinian..."

"Huh?"

"Darwinian. Survival of the fittest."

"Exactly," Steve agreed. "Most life forms vanished, but not all. The Indians migrated. The plant life, insects and animals that learned how to change, to adapt, thrived. It was life; but life of another kind."

Dawn handed Steve a sandwich and a Coke, then stood drawing a deep breath as she took in the most dramatic view she had ever seen.

"Well, you were right about one thing. Being here is like standing on the edge of the world."

1:30 p.m.

After lunch Steve and Dawn spent the next three hours road hunting, moving in the general direction of the Mid-Hills campground. By five-thirty, with their prospects for the day fading, they decided to pack it in and prepare to make camp. Far from the eighty degree heat at midday, the weather had again done an about face with temperatures dropping back into the fifties and the wind picking up steadily. Sweatshirts and jackets on once again, Steve took the truck west onto Black Canyon Road.

Midhills, so named because of its location between the

Providence and New York mountains, was maintained by the Bureau of Land Management. Said to appeal to those wanting to "get away from it all", the description seemed an exercise in understatement to Dawn as they turned onto Wildhorse Canyon Road. Still, Midhills boasted some of the most distinctive mountain ranges in the Mojave with a couple of dozen well-spaced campsites. Even more important to Dawn, there were outhouse restrooms; a convenience she had insisted upon at the outset.

Dawn's eyes scanned the sideroad foliage, stunned at the sudden and dramatic change she was observing. Within a few miles the terrain had turned from the lakebed desert to woodlands overgrown with pinyon pine and juniper.

Slowly, the camper wended its way into the crowded campground where Steve and Dawn searched for open spots.

"Damn! I can't believe it," Steve cursed. "I've been coming here for twenty years and I've never seen it full up like this."

"Well, it is," Dawn was quick to confirm, surveying the thirty or more trucks and campers parked around them. "There's not even a *bad* space available."

"Son of a bitch!" he muttered catching sight of a BLM ranger, then pulling the truck over. He rolled down his side window.

"Ranger," he called out gaining his attention. "We're down here all the way from L.A. You have anything available?"

The young bearded man ambled over to the driver's side.

"Sorry. Unusual this time of year, but we don't have a thing. Couldn't fit a bicycle in here tonight."

Steve grimaced. He'd blown it! Instead of Whiskey Pete's and road hunting, he should have gotten his ass to Midhills first thing, then proceeded from there.

"Okay, officer. Guess we're on our own."

He rolled up his window. Already he'd faced the reality of their spending another night in the camper without facilities. Now, he had to face Dawn.

Steve rubbed his palm over the side of his face thoughtfully. Inside him was an anger with himself, but deeper still a sense of resourcefulness and, yes, even adventure.

"The place where Wolfy, Ron and I saw the buck last week is about twenty miles from here out toward Tabletop Mountain. It's in a valley bound up by Tabletop and another smaller range. It's not dark yet. If we were to head out that way, it'd make a perfect campsite and we'd sure have our privacy."

Dawn guffawed.

"Privacy...My God, Steve. I already feel like we're in the middle of nowhere."

"It's up to you, but one way or the other, we're not gonna be in any campground. Not tonight anyway."

She took a deep breath.

"Well, you're the desert rat. If you think it's the best place we can find, I suppose we should go."

Dawn folded her arms tight around her, a sure sign she was annoyed, Steve noted, as they exited Midhills Campground for Black Canyon Road.

"It won't be so bad. Wolfy and I camp out like this all the time; even your brother more than once. If you look at it from the positive side, we'll be there first thing in the morning well rested and ready to go. Besides, I've got a steak and a bottle of wine that'll go pretty well together for dinner tonight."

Dawn giggled. He put his arm around her as she slid over toward him and they travelled at a forty mile per hour clip south to Black Canyon wash.

Steve slowed the truck, then took a left onto the rocky, crater ridden trail. Scattered about and separated by hundreds of acres were several ranchhouses that Steve had passed during previous trips, but other than those, there was little sign of civilization except an occasional "No Trespassing" sign posted by recalcitrant landholders.

Dawn swallowed hard as dusk settled in along with a chilling and unrelenting wind.

"Look!" Steve blurted suddenly. "See them? Out in front of us. Quail. A whole flock of them."

He stopped the camper cold in its tracks.

"What are you doing?"

Steve plucked his 12 gauge from out of the gun rack behind them.

"Quail season opened yesterday. I'm going to shoot us some dinner."

Without wasting a precious moment, he flung open the truck door jumping out to the roadside as the covey of birds ran for cover.

"Perfect," he whispered to himself carefully taking aim.

"Steve!" he heard Dawn's voice calling. "Steve!!"

His eyes diverted.

"Please don't shoot them."

"What?"

"Please. I'm asking you not to shoot them. Deer are one thing. But not them; not these quail..."

Again, he took aim, then watched through his gunsight as they passed out of range.

"If you don't beat all," he muttered returning the

shotgun to his side and re-entering the truck.

Words were few and far between as they crawled along the Black Canyon Wash to a series of unnamed trails leading to Woods Mountain.

"Are you sure you know where you're going?"

Steve just stared at her.

"I guess that means 'yes'."

He nodded.

"Well, it looks to me like the last people to go down these trails were in covered wagons."

Still, no response.

She leaned over and pecked him on the cheek.

"Thanks for not shooting those poor birds."

He grunted.

"I don't suppose I'm here to shoot quail. But I'm telling you now, if that buck comes within ten miles of me tomorrow, he's a gonner, you understand?"

"Understood," she answered unhesitantly, shuddering at the cold and prospects of treking along these uncharted backroads in the dark.

Steve flicked on his headlights as night began to fall and the radiance of a bright, crescent moon became visible above them.

"Well, it may be a little chilly, but sure clear as can be."

"Yeah, good thing, too, with these trails in the shape they are. How much farther?"

"Not far at all. As a matter of fact," he said turning the wheel sharply to the left and stopping at the lip of Woods Wash, "the place where we'll be spending the night is right down there."

He put his brights on, helping to illuminate the desert basin between Tabletop and Woods Mountains.

"It's a valley," Dawn uttered, a shiver passing through

her bloodstream like jagged particles of ice.

"Yeah, a valley. Like I told you before."

Steve did not notice the strength of emotion in her voice nor the pervasive fear she was feeling, but a queer, unspoken notion passed through her mind as they began their descent down the craggy mountainside. It came from a distant part of her subconscious framed in a scene from an old movie she'd seen years ago as a child.

To her now, it was as if they were entering the valley of the Little Big Horn.

THREE

Behind Tabletop Mountain
October 21, 1989
7:30 p.m.

Wolfy had made a lot of right decisions in his day, but buying a two wheel drive Ford pick up instead of a four wasn't one of them. Not only was the truck struggling to make it down the rocky mountainside, but Steve knew he'd need to be especially careful when he reached the canyon base where laced amid the scabrous desert floor was a wash of soft silt created by generations of rain and erosion. A trip into one of those would be one way, he was thinking as they forged their way downward, the piercing rays of his high beams sending droves of small game scurrying for cover.

"Are you sure you know where we're going?" asked Dawn suddenly.

"What's that supposed to mean?"

Steve navigated the camper through an obstacle course of cholla gardens, sagebrush and piles of granite rock.

"It means that I'm nervous about camping out here all by ourselves like this. Last night was one thing, but God Steve anything could happen to us out here!"

He pulled the truck into a well protected belt of high ground near the base of Woods Mountain. With the smaller range to the truck's back, directly facing them some five hundred yards across the desert basin stood Tabletop Mountain; its huge flattop configuration towering nearly seven thousand feet like some strangely truncated pyramid.

"Why would you be nervous? I mean, what could go

wrong?"

Steve switched off the ignition. Dawn looked straight ahead to the huge granite mountain bathed in starlight so bright you could read a book by it.

"Rapists," she answered without thinking.

Steve winced.

"You're joking."

"No, I'm not joking," she blurted, a vague sense of hysteria rising up in her voice. "There could be intruders; like a motorcycle gang or something. They could rape me and kill you and no one would ever know about it."

He started to laugh, but stopped suddenly, smitten by her chilling gaze.

"Okay, okay. I'm sorry. I know it's possible, Dawn, but I'd protect you and I swear you'd be a hundred, no a thousand times more likely to get assaulted in L.A. So what do you say? Let's set up camp, forget about all that other stuff and enjoy ourselves."

Dawn's lips contracted into a small, thin slit as if to suck in and contain the premonition of danger that had haunted her since the weekend began.

"You're right," she answered in a small voice. "I don't know what's been happening to me lately. I'm so sensitive about everything."

Steve dug a firepit in the sandy ground not twenty feet from the camper, then lined it with rocks while Dawn brewed a pot of coffee on the gas stove set atop their open tailgate. The temperature was again dropping, Steve noticed as he collected branches and juniper logs from the surrounding thickets. The campfire would be a nice touch and would keep them warm later in the night when they'd want to snuggle under a blanket roasting marshmallows and drinking wine. He smiled to himself considering that picture in his mind. As much as the buck he'd come after, that was the moment he wanted most to capture, he thought then. The two of them huddled together beside a campfire, the desert sky so clear you could see beyond the stars straight through to the very soul of the universe.

The pile of dry scrubush and wood burst into flames the moment Steve put a match to it. He admired his handiwork momentarily before making his way to the back of the camper where Dawn had already taken refuge. She sat inside, now wearing a windbreaker, sipping from a steaming cup of coffee.

Steve took a container in hand and poured himself a cup.

"Fire's started. All that's left for you to do is enjoy your coffee. I'll cook up some steak and beans. Later on, we can roast marshmallows."

Dawn smiled.

"You take care of me pretty well, don't you?"

"Damn straight," he shot back, opening a can of beans and emptying it into a pot.

"So what do you think the kids are doing now?"

Steve lay the two New York strip steaks across an oversized frying pan sizzling with butter.

"Well, what time is it?" he asked glancing at his Timex Sportsman. "Seven-thirty. I figure Bethy's asleep and Steven is just driving my parents crazy."

Dawn chuckled.

"Probably spoiling them rotten! Ever notice how hyper little Steve is after they watch him? That's because he and Wolfy sit around all day eating ice cream and chocolate together!"

"Yeah, they're like two old pals. I'm not sure whether Wolfy's a bad influence on Steve or if Steve's a bad influence on Wolfy, but they're a pair, all right." He reached inside the camper for a Coleman lantern. "Jeez, it's dark. Can hardly see what I'm doing."

Steve lit the Coleman lantern. He reached to place it atop the camper shell, then paused for an instant to comprehend the odd tingling sensation that crept like a chill up from the base of his spine. All at once and without apparent cause he felt that someone was watching. He was about to make a joke of it. Here they were, not another human being within fifteen miles and he felt they were being spied on! Funny, except that he had had that same feeling once before. But only once, he remembered as the chilling tingle plied its way through him until his entire body was left quaking in a cold sweat terror.

Cautiously, on the precipice of some nightmarish chasm, Steve glanced over his left shoulder. Relief. There was nothing. The base of the mountain. But then, as if drawn by some sixth sense, his eyes travelled up the sheer mountainside still farther, all the way to the top stopping at a sight that left him falling; tumbling into a bottomless pit of horror.

Like an atomic flash the memory of a 1975 sighting erupted. It had happened fourteen years earlier while vacationing with his parents at Lake Mojave. But in that instant the entire experience returned to him in a tidal wave of sensation: the cut on his face smarting; the pounding of his heart like a timebomb set against his ribcage; the frantic race to the safety of his parents' campsite; young Keith's panic and his own raging terror. He had seen that light; that object before. Thirty feet in diameter, pulsating its brilliance; a radiance that illuminated his most unsettling dreams and deepest subconsious fears since childhood.

He did a doubletake. He felt it and it sensed him because the instant his eyes fell upon it, the object dropped behind the mountain attempting to hide from him!

"Steve? Steve, what's the matter? The steaks! They're burning!"

He fumbled ineptly with the frying pan, his hands clumsy as tenterhooks nearly knocking it off the stovetop.

"Nothing." He turned the steaks over. "A falling star is all. I just saw it. I just was watching it now."

"You sure you're okay?"

Dawn puzzled over his sudden look of distress even as Steve worried about her. He'd say nothing of it. Pretend everything was normal. How could he do otherwise? Through some extraordinary intuition she was already petrfied of what might happen. Why else would something so harmless as little Steve's transformer bother her? And what about the morbid fear of rape and murder and the rest? No. It was his place to protect Dawn from all of what he was now feeling.

Besides, there could be explanations other than an extraterrestial craft. Maybe he'd imagined it; maybe what he had spotted was an experimental military plane from Nellis or another of the nearby bases. Certainly, the least likely of any of these was some kind of spaceship. Hadn't he spent the better part of his adult life trying to forget what he was certain he'd seen at Lake Mojave fourteen years earlier? This night as a married man and father of two, he struggled in quiet desperation to convince himself it was all some kind of mistake, then and now.

Furtively, he glanced again, then again to the mountain peak some one thousand feet above. But there was nothing further to be seen; just a sheer granite and limestone mountain blankly, stoically rising up from the desert floor.

Steve collected the steaks onto a platter, then walked to the campfire where Dawn had set two plates, two sets of eating utensils and two glasses on a blanket. He placed the platter near the center where Dawn sat huddled close to the campfire silently knowing what he had seen could not be explained away. More, that behind Woods Mountain some two hundred feet from where they camped still lurked ... something...it hadn't left and wouldn't he suspected though he could not be certain why.

He skewered a steak placing it on the plate in front of her. Next, he served the ranch beans and French bread, then sat down beside her.

"Not talking much about your buck lately. Did you forget about him?"

"No, no, I haven't forgotten. Just biding my time 'til morning when we start again."

Dawn savored her steak as Steve opened a bottle of chardonnay, then poured two glasses full.

"You know, having a meal outside like this is kind of special." She took a sip, then snuggled next to him. "I don't think I've ever seen a night so clear."

"Desert's like that. No pollution so the stars seem brighter and a lot closer. Like its own world out here."

"And noisy! I always thought the desert was quiet at night, but just listen to all these sounds."

They found themselves listening. Indeed, the night-time desert was a cacophony of dissonance, the steady buzz of insects and rustling of small game punctuated only occasionally by a coyote's piercing howl.

"Yeah, they make a racket all right and though you can't see them there are animals all around us right now. In the brush over there," he said motioning to their right near the base of the mountain, "and all through this valley."

"What kind of animals?" asked Dawn.

"Donkeys, coyotes, kangaroo rats, you name it. And don't forget bighorn sheep and mule deer. They've got to spend the night somewhere. My guess is they're here in the valley not far from us right now."

Dawn made a face.

"Rats...?"

Steve tore a piece of bread from the loaf, then tossed it near the foot of a large, mesquite tree. Within seconds, a platoon of kangaroo rats was leaping toward it, biting tiny pieces and trying to carry it away back into the sagebrush with them.

Dawn laughed for the first time since the night they'd spent at Whiskey Pete's. Even Steve managed a chuckle, the gnawing anxiety of moments before flying from him at the sight of the strange looking rodents tugging and pulling at the wad of bread.

"They're not really rats. Not the kind you'd find in the city anyway. More like squirrels except they hop instead of run...like little kangaroos!"

He took a gulp of wine, then tossed another and another piece of bread, anxious to gain even these few seconds of respite from the macabre truth he held, but could not yet bring himself to acknowledge. Together they laughed now as the kangaroo rats leapt and jumped scrambling to earn their daily bread.

"Hey," said Dawn suddenly. "How about those marshmallows?"

Steve nodded.

"You get the marshmallows from the truck; I'll get the sticks to roast them on."

Steve made his way toward the nearby mesquite tree. He pulled a pencil-thin twig from a branch. Could it be that somehow, someway he had been mistaken in what he thought he'd seen? His eyes drifted up toward the crest of Woods Mountain. He knew in his mind and heart that he had seen something; felt something...unearthly. So why now did it seem so distant? How could he doubt now what he had seen with his own eyes such a short time before?

"Steve?" Dawn called from beside the campfire.

His head jerked away from the mountaintop as if his eyes had been glued to it.

"I'll be right there," he replied regaining his composure then briskly walking toward the campfire.

He handed her a twig. She put it through the center of one of the marshmallows, then placed it over the fire. Steve followed suit.

"You must know a lot about the stars," she said looking out into the ebony sky. "What with Wolfy working at Goldstone for all these years."

"Wolfy's an electrician at the tracking station, not an astronomer, but I know a little. It's a hobby of his kinda rubbed off."

"Do you know the constellations like Ursa Minor, Orion and all those?"

He took his browned marshmallow from out of the fire; then nipped at it.

"Yeah, some. Wolfy, my Mom and me used to make a game of it. You know like out there," he said pointing across the valley over the top of Tabletop Mountain to Aries. "What constellation is that?"

Dawn studied the grouping of stars for a moment, the glow of the blazing campfire illuminating her face as she shook her head in the negative.

"I don't know. Which?"

"That's Aries, the Ram. See how the stars connect?

You've got to use your imagination. But it's there."

"Well, it doesn't look like a ram to me."

"Sometimes the stars don't look anything like what they're supposed to. In ancient times people thought they were fixed in the ceiling of a dome with humans at the center. So, they'd use constellations as symbols to remind people of some significant event or an important king or something. Like Leo, the Lion. The constellation doesn't look like a lion, but when it was named the summer sun was in that part of the sky and the heat was fierce as a lion. Get it?"

Dawn pointed to the northern sky. "Which is that?"

"I'm not sure; Pegasus, I think."

Dawn popped a marshmallow into her mouth.

"How about that?" she asked pointing beyond the horizon.

"That's Algol, the Demon Star; part of Persus. Algol is the biggest and brightest because it's a double star. If you had a telescope you could see the two of them revolving around each other."

Dawn turned to her left, eyes raising to the sky above Wood's Mountain.

"Well, if that's a double star; these must be triples 'cause they're a lot bigger and brighter than that."

Steve chortled at Dawn's sense of wonderment at it all. Starstruck as a child he was thinking as he twisted around to his left, then gazed into the horizon just above the 1,000 foot mountain. His sense of shock was physical. Like a deer he had once gutted, he could feel the cold steel of a hunting knife splaying him open from sternum to crotch.

In the sky, some two hundred yards to their right, straight up and set like diamond studs in the ceiling of a dome were nine objects shining so intensely they created daylight around them.

"What are they, Steve?" asked Dawn naively. "They can't be stars, can they?"

Steve squirmed. It was as if someone had knocked the wind out of him. His face was sheet white and his hands were shaking.

"Weather balloons," he bluffed citing the first explanation that popped into his mind.

Dawn's face twisted with skepticism.

"But, look," she exclaimed pointing excitedly, "they look like mylar. See how shiny. And they must be connected," she uttered with a sudden sense of awe, "because they're moving together like pearls on a string."

"That would fit," Steve lied in a hollow voice ringing with terror as his eyes riveted to the black ceiling sky. He could feel its presence forming like some hideously mishapen figure at the outer fringes of his consciousness. They were back, he thought, and his guts were hanging, hanging out of his wide open belly. "Maybe they're balloons; balloons connected by some kind of wire or cable."

Dawn shook her head 'no' even as she said 'yes'.

"Well, okay. You're the desert rat. But they don't look like weather balloons to me."

Steve walked to the back of the camper, eyes never diverting from the nine glowing orbs. He poured himself a cup of coffee, then walked back to the campfire where Dawn still sat sipping wine and watching.

9:13 p.m.

An eerie quiet fell over the camp. Steve contemplated loading up the truck and heading back toward Midhills, but the roughness of the surrounding terrain precluded leaving until daybreak. And that made him edgy. Every bone in his body told him there was no status quo to a situation like this; that the objects would not just freeze in the sky, then simply go away. But that is exactly what they did. There was no motion; no change in size or color. Just the interminable waiting; watching for them to make a move; for this lull before the storm to give way to a series of events whose prospect had tormented him since child-hood. Still, it was not in Steve's nature to fear or to yield and if these objects had come to do harm to Dawn or him, they would be in for a fight, he vowed silently, a battle to the death if it came to that.

To Dawn, the objects represented something else again. Unaware of Steve's earlier sighting or the maelstrom of emotions it had stirred, she viewed their presence as a curiosity; a celestial phenomenon perhaps; weather bal-loons or some experimental craft; or something far more terrifying: invasion. For weeks banner headlines had been proclaiming a mounting instability in Soviet controlled eastern Europe.   Mass demonstrations and rioting in Hungary, Poland and East Germany had created a totally destabilized world order. Could these lights; these craft that had found their way into the night sky above them have something to do with that? Literally in the middle of nowhere, how could they know what had gone on in the world during the past twenty-four hours?

Dawn sipped wine as Steve's vigilant stare remained fixed without a word passing between them, until:

"They're blinking," Dawn reported. "Blinking to one another like in some kind of communication." She turned to Steve suddenly bug-eyed. *"They're not stars and they're no weather balloons!"*

Steve didn't answer; couldn't. Instead, he studied the nine craft flashing what appeared to be coded messages one to the other, sometimes slowly and at other times together and in rapid succession.

"Could be experimental craft. Nellis Air Force Base, Twenty-nine Palms and the Marine Training Corps aren't that far from here. Could be anything."

"Maybe. But something's funny about this and I don't like it." She put down her wine glass inadvertently spilling it across the blanket. "Look at them, Steve. They're sending signals all right," she expounded, pointing now with both hands, "but they've positioned themselves directly on top of us!"

Steve sat Indian style watching, observing, his heart racing frantically beneath his stolid, expressionless exterior. The old adversary had returned. Like an unfinished match, they had come to challenge and test him. Like a predator with its quarry, they were toying with him, he was thinking as the flashing signals began to intensify. Then suddenly, without a sound and in a lapse of time so brief it seem instantaneous; the glowing objects vanished. All nine. And it was stunning.

Steve and Dawn searched the desert sky, then froze. Their mouths hung open and their bodies became suddenly rigid as if atrophied with fear and wonder. Just west of them, perhaps four hundred yards away and several thousand feet up the nine craft had repositioned themselves on the horizon just over the crest of the mountain range in the form of a large, sprawling 'M'.

"Steve, I'm scared. Things like this aren't supposed to happen even if it is the military." Dawn became suddenly animated. "Maybe we shouldn't be seeing this; maybe they're doing some kind of secret exercises. I'm afraid...and don't ask me why...but I'm afraid they know we're here *and we're somehow a part of all of this!*"

"I don't know," Steve answered, "but I've got a gun; two of them. If it's the military they may know we're here, but won't bother us..."

Dawn reached for him. She held his face between the flat of both her hands.

"You said *if* it's the military like you don't think it is. If it's not the military, *our* military; what could it be?"

"A lot of things, I guess. The simplest explanation is military testing. I've read about them doing night time exercises; desert training and the like at Twenty-Nine Palms." He stared deep into her eyes. "But whatever it is, please don't be frightened. Hey! They're on our side, remember? It's people like you and me that pay their goddamn salaries!"

He gave her a look of reassurance and a nervous laugh, then looked up from the valley basin to the sky once again.

"*Holy shit,*" he whispered.

Dawn grabbed his upper arm and squeezed it so tightly she could feel the bone.

"*My God,*" she uttered in disbelief.

The sky above them; their entire nighttime sky was now filled with *hundreds* of shiny objects flashing signals to one another!

Together they sat holding each other, too frightened to move. Too frightened until it began. Slowly at first. Like parachutes floating, free falling downward so gracefully, so easily that it seemed indiscernable at first. So they stared; captivated by the prospects as dozens; no hundreds of the glowing, white objects travelled like falling stars down and across the black, onyx sky. Subtle, graceful, beautiful in their way, the round and shining orbs were descending upon them. *The valley was, in fact, being invaded!*

"Oh, my God!" screamed Dawn. "It's the Russians! The Russians are invading!"

"No, no," Steve comforted, scrambling to his feet. "It can't be! It fucking can't be!" he repeated, then began kicking out the campfire.

"Get in the truck!" he screamed to Dawn finally. "I'll put out the fire so they won't see us!" She hesitated; he shoved her. "*Get in the Goddamnn truck!*"

The lights were landing now, Steve observed, landing all over the valley. The lights were landing, his frantic mind processed, and coming toward them!

With the campfire decimated, Steve grabbed Dawn who just sat there staring and jerked her to her feet.

"They're coming to get us," she muttered, entranced.

"There's no road. No roads out there," Steve said as much to himself as to her, "but they're still coming at us!"

He looked to Dawn who had lapsed into a state of shock, then took her by the hand as he raced around to the truck's cab grabbing his 12 gauge Ithaca shotgun, 7 MM Browning rifle and a hunting vest looped with a dozen rounds of ammo.

Steve nudged Dawn away toward the back of the camper shell. "Go! Go! Get in the back where it's safe," he screamed, lodging two shells in the shotgun's magazine and one in the firing chamber. But Dawn would not or could not leave him, though the sense of urgency jolting within her was electric.

Unbelievably, the bright, glowing objects were continuing to land; dozens and dozens, hundreds and hundreds raining down from the now blackened dome-like sky, hitting the desert basin then bolting forward in a steady fast-clipped progression toward the camper.

"Come on you, motherfuckers! Get within my range!" Steve threatened taking a rifleman's position as Dawn clung to his arm attempting to pull him away, back into the camper shell with her.

"You can't do anything. You can't hurt them! There are too many of them and they're too advanced; too advanced for us!"

Steve looked to her as if emerging from a dream, suddenly aware of her tugging at his arm.

"*Don't you hear it? Don't you hear what they're telling you?*" she was screaming.

He stared down at her for a split second; not understanding.

"We can make a run for it. Get in the truck and haul ass outta here."

"No, no you can't. There are no roads. Look at them rushing toward us! They'd kill you and rape me! Don't you understand? You were right the first time. It's the camper; the back of the camper where they want us! Come! Come on!" she pleaded pulling at his arm.

Confused, and convinced they had no chance of making it out of the valley, Steve followed her, to the back of the camper, then inside.

Quietly, Dawn sat watching as Steve pulled up the tailgate with the window down laying the 12 gauge barrel over its metal edge then taking aim. Nearly all of the shimmering white lights had fallen from the black void that that been created and were now rushing toward them.

"Come on you bastards," Steve called out in a voice that echoed across the valley, now absolutely devoid of all sound, animal, human or otherwise. "Come on! Come and get it!" he was ranting, all of the pent-up fears and emotions blaring, blasting like a volcanic eruption from out of him.

Then, Dawn grabbed him by the arm, *hard.* "Don't. Don't do anything to hurt them." And again, the Voice. *'Don't do anything to hurt us.'* And now, Dawn pleading. "Put the gun down. You have no chance." Then the Voice, coldly. *'Put the gun down. You have no chance.'* And Dawn panicked. "In the end, you'll only get us killed if you try to harm them." Them, threatening, *'In the end, we'll kill you if you try to harm us!'*

Steve turned toward her angry and disoriented.

"What are you trying to do? Whose side are you on, anyway?"

"Look," she uttered, pointing index finger extended; arm stretched out full length.

Steve recognized the glint of terror in Dawn's eyes, as he swung around to defend them. But there was nothing to defend against for standing just beyond the back of the camper were two... Beings. The sight of them; the smell; the detail of the physical characteristics are something he will never be able to forget. They appeared three feet in height and two feet in width, though their proportions seemed not yet to have taken total physical form, both expanding and contracting alternately. Bluish-gray in color of the type caused by two live and crackling electrical wires; it was as if they were holograms; physical and yet not physical; translucent and ill-defined but unquestionably present.

He recoiled in horror, then fell forward, bracing himself with his right hand. And in that moment, they rushed toward him to an equal degree. He regained his balance sitting up and was stunned to observe the two gray entities drop back again, incredibly, to the same equal degree. It was then that Steve realized they were stationed there as guards to monitor their actions and make certain they didn't get away.

"You see it's hopeless to fight," Dawn tried to reason. "They're not military, at least not our military."

And she was right. The excitement, the voice, the dizzying emotions, were overwhelming. Without thinking, Steve lay the shotgun on the camper bed at his side as Dawn reached for him, wrenched with a numbing, wracking fear. Steve had lost his will. So unlike him. So outside his realm of experience. He sat still, blindly stroking the side of Dawn's face with his hand, his eyes fixed upon the two alien beings set as guards before them; then stopped. He heard Dawn gasp, then felt his own breathing cease as if the air had been sucked out of his lungs.

The brilliant white objects had all landed now and the sky was chillingly empty of light; devoid entirely of either stars or moon. But the valley, dark and sprawling; coralled by the shadowy outlines of the enormous mountainscapes, was filled; overrun from several feet away to hundreds of yards beyond with *thousands of pairs of red eyes glowing in the dark around them.*

II. MIND TEAR

They wanted everything we had...**everything**...our minds, our bodies, even our souls, I think. It was like they drew it out of us with a syringe...every molecule. And it was painful and I thought we were going to die, or had already died and were being tortured in Hell.

Dawn Hess

ONE

Mojave Campsite
October 21, 1989
10:55 p.m.

What is happening to us? Steve and Dawn anguished as they stared from out of their camper shell literally pinching one another as thousands of fierce and threatening pairs of red glowing eyes surrounded them.

The sky was moonless; the valley so terrifyingly soundless that the wind had stopped entirely; not an insect chirped; not a donkey bayed. The remnants of French bread and marshmallows they'd left by the campfire were laying there untouched and any sign of animal life - even the kangaroo rats - was gone as if it had vanished from the face of the earth.

Dawn stared out from the back of the camper where the two electric, gray Beings stood as monitors. Then to her left and right where through the windows of the camper shell she watched the eyes move close enough so that the shape of their dwarfish bodies became apparent.

Like gremlins, no larger than three feet in height, with heads the size of a cat's, translucent torsos and thin diaphonous limbs, the creatures revelled around the camper with boundless energy. Their eyes, fierce and malevolent, pierced the night as a pack of them climbed the branches of a nearby mesquite tree while others tumbled and frolicked like child-monsters amid the sagebrush and juniper that shrouded the desert wash where they'd camped.

"We're going to die," said Dawn at last.

These were the first words uttered by either of them since they had positioned themselves in the back of the truck. Words seemed so inadequate, they thought then, so incapable of carrying the emotions they were feeling. Besides, there was something else at work now, something deeper going on, they began to realize, as the Voice inside their minds anchored itself ever more deeply so that beyond influence, it was exerting control over them.

Steve's wary eyes scanned the blackened valley, unsure of himself and all that was going on around him. His huge frame quaked as he reached for Dawn's hand.

"Pinch me."

"What?"

"Pinch me."

She did. And he felt it.

"Am I awake?"

"Yes."

"Are you seeing what I am?"

"I think so, but explain it to me," she answered numbly. "Tell me exactly what you're seeing so we can compare."

He swallowed hard. It was an exercise in sanity.

"The monitors...grayish-blue...like electric images...forming and unforming...at the foot of the tailgate."

She put her trembling arm around him, pressing him tightly to her.

"Yes. That's what I see. What else?"

"The eyes..."

"Red?"

"Yes."

"What do they look like?"

"Dwarves...kind of like monkeys, but evil. I can feel the evil. See it in those eyes." He shook his head in disbelief, attempting to erase the image from his mind. But it was no use. "They're playing," he continued. "Like wild kids;

running up and around the camper into the wooded area and brush of the foothills and in the tree over there," he said pointing to their right. "But they must be... weightless."

"Why?"

"Because the branches; the limbs of the mesquite, even the thinnest ones, aren't bending."

"I know," Dawn whispered. "I see it, too."

Steve held her at arms length suddenly. He looked deep into her eyes, burning with fear.

"Are we hallucinating?"

"No. I don't think so."

"Then, it's real?"

"As real as anything I know," she pledged. "Steve, I'm scared." She put his hand over her heart. "Can you feel it? My breathing...my heart...I think I'm hyperventilating."

Steve's jaw locked. He cast a lethal stare out to the gremlims whose sole purpose seemed to be to mock their horror.

"Don't panic," he told her, a jolt of adrenalin cascading through him. "They must want something and I'm going to find out what it is."

He attempted to rise. The monitors rushed forward before Dawn could restrain him.

"What do you want?" He shouted out to them.

No response.

What the hell are you doing to us? We have a right to know!" he demanded. *"We're human beings. We have a right to know!"*

Still nothing. Only the monitors' expressionless presence, vigilant, without emotion, and positioned now at the very lip of the tailgate.

Steve removed Dawn's tempering hand from his shoulder, sliding forward still further. The Beings rushed toward him to an equal degree.

He reached out to touch the one nearest to him with his right hand, then recoiled as the burning sensation of an electrical shock ran up from his fingertips through his outstretched arm.

Steve was jolted backward. Dawn rushed to his aid. She spoke, her voice suddenly desperate.

"Please, don't do that again. Please don't ever leave me, Steve. Promise. Promise me now."

He rubbed the stinging fingers of his right hand; their eyes met. Dawn's face was ashen. He could see the helplessness in her eyes.

"I won't leave you, Dawn. I'll never leave you," he vowed pulling her closer to him. "I love you more than anything. Even my own life."

Together they held one another sobbing as the gremlins frolicked with reckless abandon seemingly stimulated by the wild rhythms of their elevated heartbeats pounding chest against chest. Then, Steve's body turned rigid.

"What is it?" she asked. "What's wrong?"

She looked up to him. He stared over her shoulder wordlessly. His glistening eyes were fixed and dazed.

"Are you all right?" she fretted turning her head to see...It. The spacecraft. Huge. Descending from out of a dense cloud. The object literally capped the valley, then stopped, hovering perhaps one hundred yards above the desert floor.

The craft was larger than anything they ever imagined could fly. More than a football field in diameter and shaped like a disk with an elevated dome that rose up from its center, it was encircled by brilliant white lights flashing in what appeared to be coordinated, coded rhythms.

Together they watched in awe, stunned as the huge

craft shot a probe perhaps two hundred feet in diameter, down to the ground, then began transporting objects both into and out of the spacecraft. From the disk's underbelly hung six smaller units the size of helicopters and designed like miniature versions of the mothercraft that suddenly became visible. All of this happening right before their eyes. And it was overwhelming.

"Do you see it, too?" Steve asked at last.

"Yes."

"The big ship with six others suspended from it?"

She nodded, her mouth buried in the flesh of his shoulder.

"The tunnel of light beaming objects up and down?"

Again, she nodded.

"Name the objects..."

"They're patterns of things more than actual physical objects. Things like plants and cattle; cactus and even trees."

"Anything else?"

Dawn paused long enough to observe a pattern of red, amber and white lights in the form of a triangle emanating from the ship, then breaking loose from it; moving carefully, methodically across the desert basin.

She nodded. "Yes. There's a triangle of lights with the tip of the triangle touching down on the ground. It seems to be combing the ground beneath the disk as if searching for something." She shook her head to clear it. "Do you hear the rumbling? Deep inside the ground. The triangle is like a drill; a huge drill of some kind."

Steve looked out into the horizon where the inverted triangle would stop momentarily, then continue its methodical search. He stared into the red, amber and white triangle, his eyes totally unaffected by what would normally be a blinding radiance. The sound, too, that Dawn alluded to could also be heard by him. A low, hydraulic rumble that seemed to reverberate up from the very bowels of the earth.

It was then that he became aware of the smell. A rank, chemical odor like phosphorous or burning rubber. It was sulphur, Steve decided, just before his heart stopped and the words issued forth from his mouth in the lifeless, monotone of saturate shock.

"What else?" he asked her.

"Huh?"

"What else do you see?"

"Jesus..." Dawn sobbed. "...My dear, sweet Jesus," she repeated, then broke into prayer. "Dear Heavenly Father, please forgive us all of our sins. Take us home and don't let us suffer. Provide for our children who have done no one any harm. Protect them. Save them from this evil," she implored, then began crying, for peering into the camper window, totally illuminated, was a third type of Being.

11:30 p.m.

Steve knew immediately that these were the ones responsible for the telepathic messages. They arrived in a group of nine circling the camper, staring in the windows of its shell like spectators observing animals in a zoo. Taller than the others, they stood perhaps five feet in height, with enormous raven-black eyes and long spindly appendages connected to an underdeveloped torso the size of a three year old child's.

Dawn stared into the face of one of them, if only for a second, but it was long enough to brand a searing image that will stay in her mind forever.

"It's the children they want," Dawn swore with deadly conviction. "It's Steven they're after."

"Look at them," snarled Steve, his mind tracking in a totally different direction. "They're scientists all right. Trying to get inside my brain. Studying me. Observing me like some kind of bacteria under a microscope."

"They're holding us hostage now while the others go after him," Dawn continued. "Little Steven is calling out to me in his dreams. I can hear him. But I won't let you have him because I know what's it's like to be taken away and treated like some kind of thing. And it hurts," said Dawn sobbing. "It hurts so deep and so long that you never forget it!"

Steve glared out through the camper windows as the illuminated figures continued their grisly parade before them. He remembered his football days at the U. of R., his prowess as a hunter and all that made him who he was, and it angered him to feel so helpless now; to know that their lives and the lives of their children were in danger and to be so impotent to do anything about it.

"Damn you bastards! You think you can just hold us against our will," he exploded. "We're not animals you can put in a cage and perform experiments on. We're people. People with thoughts and ideas and souls!"

He sat up straight and confrontational; the blood rushing to his head; his rage mounting. And it was then that it hit him. The flash in his mind. WHAM! It was blinding. He reeled backward. Was he having a stroke?

"Steve, what is it?" Dawn asked, moving toward him. "Are you all right?"

He stared deep into her eyes, wary and frightened. His sense of helplessness was palpable.

"I, I don't know. I just felt something. It shot through

my head like..."

But before he could finish, the visions began. Strange, unconnected pieces of his life history and emotions; fractured, they spun dizzying through his mind like shards of a shattered mirror.

Steve is in Bulldog stadium playing the East-West championship bathing in the cheers of thousands as teammates give him hugs and high-fives after a game-winning touchdown...

"Do you feel it?" he asked.

Dawn paused to gain a sense of what might be happening.

"No. I don't feel anything."

Steve stands, a seven year old boy, fishing Crestline Lake with his Dad and brother, Ron, reeling in not one, but two bass on the same line...

"What does it feel like?" Dawn puzzles. "What are you feeling now?"

Steve looked around the camper, disoriented, to Dawn, then to the bed cluttered with gear and finally to the truck windows brimming with the brilliance of the illuminated figures.

"Memories. But more than memories. Feelings. I'm feeling what it was like seeing and doing things; not just remembering."

Steve is hunting the high desert with a cadre of pals during his senior year of college. They flush the thick brush finally coming upon a covey of quail. He shivers with delight as the birds take to the sky, the sound of shotguns blasting in his ears...

"No. Not painful. Pleasant, mostly," he uttered, stopping mid-sentence.

He is a child in the living room of his parents' Barstow home unwrapping presents beneath the tree on Christmas morning...He is a college student, heart aflutter seeing Dawn across the Quad at U. of R. for the first time...He is lifting weights at the spa...He is on his knees proposing to Dawn at the Top of the Tram restaurant in Palm Springs...

"What's going on?" He heard Dawn scream, suddenly hysterical. "What are they doing to you?"

Steve is exultant...Steve is humiliated...Steve is proud...and jealous...and uproarious with laughter...

"Steve? Steve!!" Dawn shrieked in desperation, but it was no use.

Steve tried to answer, but couldn't. Nor could he hear Dawn's voice finally crying out for him as he stared, mesmerized, into the faces of the aliens. Those fucking faces; so detached and unemotional, seizing him; transporting him to a place where his worst fears and most elusive nightmares crystallized into reality.

Fifteen year old Steve Hess looks back toward the orange-red glow of the campfire where his parents sit drinking coffee with Keith's folks, the Hunters. His family had been coming to Lake Mojave ever since he could remember; but this time is different. He knows it, though he can't say why.

His high-top Keds shuffle through the gritty soil as the adult voices evaporate into the chill night air. He takes a mental accounting of how they got here: Ten miles beyond Searchlight, park the car at North Cottonwood docks, trek further north three miles beyond the foothills to this the rocky banks of Lake Mojave. Here they are. Same as one dozen times before. So why is he suddenly so frightened?

Steve shoots a furtive glance at Keith, who is three years his junior. He seems unfazed, fishing rod angled over his right shoulder like a soldier as he walks.

"You like fishing at night?"

"If you want to catch fish, you gotta get 'em when they're hungry," Steve answers trying to shake the sudden iciness that begins to take hold of him. "There's two times when they're hungry. Early in the morning and late at night."

Keith nods wordlessly. The two boys continue down the dirt path finally emerging at the base of a large, flat rock near the lake's edge where they set down their gear.

Steve takes a Coleman lantern from out of his knapsack, then lights it as Keith baits his hook. He places the glowing lantern between them, his eyes slowly raising to the stark Nevada mountain range across the lake some four miles away. Above them the full moon is blazing. Off to the left the eerie glow of Las Vegas lights reflects like a nightfire on the horizon. And it's then that it hits him. The feeling. It's as if they're being watched!

Keith casts his line into the water as Steve baits up and does the same.

"Hey? What are we fishin' for anyway?" Keith finally asks.

"Catfish, bass—anything we can catch," the answer comes back.

Hooks baited, lines in the water, they wait.

"Fishing's a matter of patience," Steve explains. "Patience and...Jesus! You've got one!"

Keith fumbles the rod between one hand and the other. Steve takes a step toward him to help.

"Hook it! That's it!" he urges as a three pound bass breaks water not fifteen feet from the shore. "He's swallowed the bait, Keith. Now. Snap the line up!"

Steve puts his left hand over Keith's quaking grip helping to steady the rod when suddenly he feels the hairs on the back of his neck raise.

He sees it first in the recesses of his mind even before his eyes lift to the top of the mountain range. The large, shiny orb stares down at them with the magnetic engagement of a human eye, staring; watching. The size of a basketball even from across the lake, it rises up from behind the mountains, perhaps five hundred yards, then sets itself stationary, boldly, for anyone to see. But it isn't anyone, Steve realizes immediately. It's them!

His hands drop to his sides reflexively. The pole jerks from out of Keith's hands, then scrapes a trail along the coarse, sandy bank like the fingers of a dying man.

"The fish! The fish! It's getting away!" Keith screams. He swings angrily toward Steve, then falls silent.

Stunned for the moment, the two boys watch, mesmerized, as the blinding white light darts from its position due east to one directly due north before their minds can register what their eyes have seen.

"What, what is it?" the younger boy whispers hoarsely.

Steve shudders.

"I don't know, but it's seen us, man. It's seen us. Look at the way it just hangs there starin'. And listen..."

Keith drifts spellbound toward the edge of the rock. "I don't hear nothin'..."

"That's right. Not a bird or an insect. Not a rat or coyote. Like the whole world got turned off."

Steve studies the brilliant white light. Beside him stands Keith tumbling mind and soul into the abyss of terror that had been created.

"I'm scared, Steve..."

Steve's eyes grow wide. His heart pounds as he begins reeling...reeling in his line.

"Pack your gear."

"What?"

"Pack your gear now! It's coming toward us, Keith. Jesus Christ! It's coming to get us!"

Steve snuffs the lantern. They leap down from the huge flat rock, Steve shoving Keith in front of him, looking over his shoulder as they run. From simply moving, the round shining object seems to turn oblong like an egg; its pulsating white light gaining in intensity as it bolts forward.

"Run! Run!" Steve exhorts. "Please, God! Please, God!" he begs breathlessly. The object is nearly on top of them!

Steve pushes at Keith's back. He is puffing, near spastic with fear, when Steve overtakes him, looping his thick arm around his waist, sweeping him from the ground. But it is a mistake. Coming as close as he does, their legs tangle and both fall face first into the sandy earth.

Keith is trembling. Steve feels a thick, choking sensation that leaves him paralyzed, barely able to breathe.

Slowly, he cranes his head around, then peers up over his right shoulder as Keith staggers to his feet - running wildly back toward the campsite.

Directly above him, the craft hovers, a shaft of white light probing like a laser down onto him.

"What is it you want?" he whispers hoarsely. "What is it you want from me?"

Steve's body became a shivering mass in Dawn's arms as tears streamed down the sides of his face. His eyes flashed open momentarily praying it was all some kind of horrible dream. But it wasn't a dream; not even a nightmare.

He looked outside the camper windows covering his eyes at the sight of the illuminated figures, still glaring at him, understanding for the first time the depth of what was happening. More than an object, Steve was convinced, the craft he had spied over Woods Mountain was the object; the same he had encountered more than fourteen years earlier while vacationing with his parents at Lake Mojave!

The realization swept through him like a chilling wind. It seemed impossible. But it must be; had to be true for he was living it.

For whatever reason, they wanted him; to kidnap; to torture; perhaps even to kill and he was absolutely powerless to stop it.

11:35 p.m.

A feeling of total helplessness gripped Dawn as she stared into the faces of the illuminated figures. Like a computer on overload, she was taking in too much... viscerally...psychologically...so that she was certain she had crashed through the barriers of normal tolerance and was now insane.

"Steve?" she rasped.

He swung around, the thought of his wife in danger jarring him back to his senses.

She appeared distant, dazed. Her facial expression was blank; her eyes glistened.

"Dawn, what is it? Are they inside your head?"

She nodded.

Dawn is a small girl listening to Haydn's E-Flat Major concerto blaring from their living room sound system, snuggled up to Dad on a Sunday morning...

"Dawn, you've got to listen! The visions I told you about. I was wrong. They're not good. You've got to fight them! Got to keep them from gaining control!"

She is at home plate swinging a bat and hitting the ball in her first Little League game, Mom and Dad cheering from the stands as she rounds the bases...

Her eyes shot frantically to the enormous spacecraft hovering above. Then, to the gremlins circling the camper, threatening to overrun it. And finally to the tunnel of light sucking up entities of every size, shape and form as the ship's triangular "searcher" continued scouring the desert.

"Don't let them take me, Steve! I don't want to leave you!"

"Then don't!" he urged, grabbing her at the shoulders, attempting to shake her back to reality. "You can control your thoughts. You can control your mind!"

She is an adult in the familiar environs of their local church in Upland standing next to her mom singing with the congregation...She is at a party celebrating her sixth birthday, surrounded by friends and family, blowing out the candles on a cake...

Dawn began to pray, desperate; certain she and Steve were going to be killed.

She is making love with Steve...She is giving birth to Bethany...She is hearing the news that her boyfriend Dave Horton has cancer...

Dawn pondered her God and faith as a Mormon, begging that the children be allowed to understand that they haven't been deserted; that she loved them now and always would.

She is playing flute in the college orchestra...She is learning of little Steven's heart defect...She is shopping the French Riviera with her mom...

And then it came. The flash in her mind. BANG! A jolt so pervasive that it swept her away; plunged her into a nightmare reality from which there was no turning back.

The hospital room is cold, almost glacial, with every chair and table; every object composed of stainless steel or formica. The fluorescent lights above are huge and radiant; devoid of either warmth or comfort. And it makes Dawn nervous.

She feels Steve's hand on hers. Her eyes raise to see him standing over the bed, puffing and blowing, "coaching" in the style of LaMaze, but she is afraid events have taken them beyond that.

"Omigod!" Dawn screams, gripping the sheets with both hands.

"You're doing fine, Dawn," Doctor Wallach says calmly. "Steve, let's see how we're coming along with this baby."

The nurse lifts the sheets covering her legs. Dawn feels fingers going back inside her. They already told her the baby is transverse. But she could never imagine this. The pain is excruciating; unrelenting and after nearly sixteen hours of labor she is uncertain how much more she can take.

"Please, Doctor! Please, help me! Please do something!"

"Nurse, 50 mg. Demoral; 50 mg. Phenergen IV." He smiles gently. "Don't worry. We'll take care of you, Little Missy," he soothes. Then, turning to Doctor Sublett, the Neonatologist, "Contractions are one minute apart, but we seem to have stopped dilating at 5 centimeters."

Doctor Sublett glances in the direction of the NST monitor.

"The pitocin should be taking effect momentarily, Bob. Let's give it a chance."

"Right you are, Doctor."

Dr. Wallach straightens up, then snaps off his surgical gloves, replacing them with a new sterile pair. He places his hands on Dawn's lower abdomen attempting to reposition the fetus into a crown-first postion, but it's just not happening.

"Nurse, any change in the progress of labor?"

"Yes, Doctor," she answers, gently wiping Dawn's forehead with a damp, cool washcloth. "She's at seven centimeters and seventy-five percent effaced with contractions still one minute apart."

"Good. Very good," says the doctor. "Your cervix is almost entirely dilated."

"Steve!" Dawn screams. "Ste-eve!"

His expression is distressed as he moves toward her.

"I love you, Dawn," he says; the first thought to cross his mind.

Doctor Wallach has positioned himself between her legs.

"You're baby's ready, Dawn," he says working the fetus into position.

"Push," he urges. She is straining. "That's it. Good. Very good, Dawn. Push harder now."

"Go on, Dawn. You can do it," Steve exhorts.

She gathers her breath, then forces the oxygen down with all of her strength, expanding every muscle in her body.

That's it. I've got a shoulder; and now an arm. Come on, Dawn. Don't stop now. Keep pushing!"

Moments later, Dr. Wallach pings the soles of the infant's feet with his forefinger. The baby responds with an energetic wail. He clamps and cuts the umbilical cord, then hands the newborn over to the waiting resident.

"Congratulations, Dawn and Steve. You are now the parents of a baby boy."

"All right!" Steve exclaims triumphantly. "Steven Ray Hess, Jr.!"

He kisses Dawn on the forehead.

The resident physician uses a bulb syringe to suction the child's nose before the nurse joins him to record the Apgar scores.

"Apgar of four. But I don't like the sound of his breathing," says the resident physician casting a worried glance in Dr. Wallach's direction.

The obstetrician rushes over to him. They confer.

"Nurse. Get this little guy into neonatal stat."

*   *   *

Hours have gone by since Dawn plunged into a deep and dreamless slumber. The lights in the hospital room are on; glaring. She opens her eyes. They adjust.

"Where's my baby?" she asks.

A nurse, the only person in the room, seems nervous. The reaction puts Dawn immediately on guard.

"You just wait here while I get the Doctor."

She depresses the "Call" button putting the nurses' station on alert.

"Where is my baby?" Dawn asks again; this time more emphatic.

The nurse edges toward the door.

"Now you just stay calm. I'll be right back with Doctor Wallach."

Through the grogginess and headache, Dawn hears a baby crying. She fights the drug induced lethargy, then struggles out of bed. She slides her feet into slippers, and puts on the pink and white housecoat she had brought with her to the hospital as the Doctor and nurse enter.

"You can't get out of bed," the nurse protests rushing to her. "You're too weak."

"I can and I will," she answers bravely. "Where is my baby?"

"Dawn, just relax now and let me fill you in." Doctor Wallach pauses. His voice is soothing. "Steven was born with something known as ventricular septum defect. Now it's nothing life threatening and nothing to be overly concerned about." She gasps. "Steven is in the neonatal unit and being given the very best care possible."

"I want to see him! I want to see him now!"

She shuffles toward the door. The nurse moves to restrain her. Seeing her determination, the Doctor raises a palm in the air signalling 'let her go' as Dawn opens the door, then leaves the room.

Still confused and disoriented, the effect is dreamlike, but the urge to move toward the cry becomes stronger as she passes a covey of physicians; seems to float by them.

"Sad. Quite sad," she hears one of them comment.

"A goddamn tragedy is more like it," says another.

But she pays no attention, the sound of her baby's cry far more compelling than any conversation.

Then, there is Steve. 'Don't you hear him crying? How can you be standing here with Steve, Jr. crying like that?' she feels like asking, but says nothing as they pass in the long, white corridor.

"They tried. Did everything they could," he mutters absently. Is he talking to her? "But they were outnumbered." He stares straight ahead at her. "There were five of them...five with gizmos like they'd never imagined. He shakes his head confounded at the magnitude of the situation. "I'm...I'm sorry."

Dawn is too numb to react. If she can concentrate on only one thing, she tells herself, it will be the cry. 'Don't worry, Little Steven. I won't let them hurt you,' she promises, still pressing forward steadily, one foot at a time, when she encounters Wolfy and Diane sitting in a waiting area outside the operating room.

Wolfy looks to her plaintively, eyes brimming, as she passes.

"We did everything we could, baby. Everything. There was just no way to know it would turn out like this...And they have their ways. Believe me. They have their ways!"

Dawn moves forward. The infant's cry like a hook planted firmly in her brain pulling, drawing her toward it.

"We love you, Darlin'," Diane calls out after her. "No matter what. We love you. And you know he'll adjust...kids always do."

Dawn doesn't react. Can't. Instead she gravitates toward a set of panel doors. Above them, in bright red block letters, read the words, 'OPERAT-ING ROOM'. She enters. Her eyes blur momentarily as she attempts to focus.

Positioned midway across from her in the chamber is an operating table. A large circular light hangs above and it is from beneath this light that the cry emanates.

Perplexed by what she sees, she must have walked in during the middle of an operation, she is thinking, because surrounding the table are four physicians. They turn as she enters and she sees that the cries come not from an infant, but from Steve, Jr., their three year old.

He sits up. His chest cavity is open with a pulmonary clamp attached to his circumflex artery and a catheter fixed to his left atrium. An IV unit meters plasma into his right arm as blood pumps freely from an open artery leading to his superior vena cava.

"Why did you leave me, Mommy?" he asks staring straight ahead at her. "You left me alone and now they've taken me with them."

"No. No, it's not true. I would never leave you," she pledges. "Never! Ever! Leave you!"

She rushes toward him, then stops dead in her tracks. Slowly, her hand raises to her mouth; overcome by the horror of what she is seeing.

These are no doctors, she realizes, as the sulphurous odor permeates the air around her and the eyes become more prominent. It is Them.

"Mommy! Mommy!" Steve, Jr. shrieks, the life issuing from him as the Aliens' chilling stares bear down more forcefully upon her. "They've taken me. Used me for experiments!"

"No!" Dawn screams twisting and turning to break from their grip. "No! No! No!"

But it was not the creatures who held her now. It was Steve. They were in the camper again. And they were surrounded. Held captive as the illuminated figures turned their world inside out.

At first Dawn thought she had gone insane or that she and Steve had died and were being tortured in Hell. Whatever the explanation, she was made to feel a sense of loss so devastating that she wasn't sure she could go on living.

Dawn was coming apart. They were coming apart, Steve had the presence of mind to realize. Her entire body was quaking and so was his. Their hearts were pounding erratically and their body temperatures fluctuated wildly: elevating in a smothering welter that stole the breath from their lungs one moment; shivering uncontrollably with chills that cut through them like the cold steel of a razor the next.

Psychologically it was no better. Neither seemed capable of coherent logic. Sentences were becoming difficult and during some intervals impossible to put together. Their mental state was predicated now not upon physical input interpreted and processed, but on emotions and reactions already processed, then implanted into their minds, one after another, until they could bear no more.

Dawn looked to him, her face sallow and bloodless.

She was having difficulty; grave difficulty breathing.

"I don't think I'm going to make it, Steve," she confessed in an ardent voice. "I can't breathe. It's like there's no air and I'm suffocating and about to...die."

The streams of tears made their way down her cheeks. Steve kissed them.

"I feel it, too," he uttered weakly. "Let's decide now that if we die," he gasped, "and if we can, we'll stay together forever."

She nodded. He looked a final time to their captors, himself on the verge of collapse. Like scientists who had just dissected a laboratory animal, the illuminated figures observed totally removed; totally detached. Not a sign of emotion or pity could be discerned; no remorse, not even curiosity.

The stench of sulphur hung in the air. A vague sensation of nausea passed through their minds as they felt themselves slipping away into unconciousness. But it was not the dark numbness of death that followed. Rather, there came a feeling; subtle at first, that hinted they were still alive as a pervasive chill entered the camper and began to revive them.

The vapor rolled into the shell from behind the monitors like a frothy mist easing both their physical and psychological anguish. Within seconds, their pattern of breathing began to alter as each felt a pressing weight on their chest that regulated their intake of oxygen. It was controlled now. Better. More regular. But beyond their influence as if the creatures knew they were dying and wanted to keep them alive.

But why? For what purpose?

TWO

Mojave Campsite
October 22, 1989
12:15 a.m.

The chilling mist that had entered and filled the camper slowly receded leaving both Steve and Dawn psychologically ravaged.

"What happened?" Steve asked, just now coming to his senses. "Where were we?"

Dawn did not attempt an answer, but turned from the camper windows physically revolted.

"*For God's sake, pull the curtains closed!*"

He stared at her.

"*The curtains, Steve!*" she repeated, deflecting the creature's radiance with her hands. "*I- just - don't - want - to - see - them!*"

Steve pulled them shut forcefully. It did no good. Their bodies. The eyes of the illuminated figures shone through the flimsy nylon as prominent as ever.

He held her.

"What are they doing to us, Dawn?"

She shook her head at the incomprehensibility of it all.

"I don't understand, but I know it has something to do with Steven. It was like a dream, but absolutely *real*. I was there in the hospital with him and it was *horrible*! Worse than an operation. It was like some kind of experiment...some kind of dissection!"

Dawn began weeping. Tiny, anguished cries at first, followed by prolonged sobs that rose up from the very depths of her being.

"They want him, Steve. They want our son. I don't know why...can't imagine...can't believe they'd want to hurt him. He's innocent," she pled in desperation. "He's never done anything to anyone. So, why? Why would they want to do what they did to him?"

"It's all right. It's going to be all right," Steve comforted, his eyes raising to the creatures outside the camper once more.

"Mine wasn't about Steven," he said slowly. "It was about me...and them. I never told you. Never told anyone because I figured you'd think I was crazy. But when I was a kid, fourteen years ago, I went fishing with a friend at Lake Mojave. It was night and my folks were at a campsite about a quarter mile away when we spotted a bright light over the lake." He sniggered. "Wasn't hard to spot, hovering over the mountain range and clear as it was that night. And it followed us. Chased both Keith and me until neither of us could run any farther."

"I remember we both fell to the ground. I stayed right there in the dirt, paralyzed, unable to move a muscle while Keith got up and ran away. Then, it just floated above me. Without a sound, the UFO just hovered over me shining this light. A white light so bright I'll never forget it, but one you could stare right into without blinking; without it hurting your eyes at all."

Steve glanced to Dawn, the years suddenly peeling away to reveal that same terrified adolescent, more vulnerable than she had ever seen him.

"I don't know what happened after that, but early the next morning I found myself back at the campsite. Keith was asleep so I didn't wake him. But hours later when he awakened he didn't remember a thing. Not the sighting. Or even us running from it like we did. Nothing. So, I buried it. A secret I've never told anyone, until now. Until a moment ago when the entire experience came tumbling

back out of my head like it was happening all over again."
He let loose a sob, then caught himself. "And they've been
waiting for me ever since, Dawn. I believe that, though I
don't know why," he confided, tears suddenly streaming
down the sides of his face. "I believe that they saw me then
and have come back for me and you and our kids, now,
after all these years. But I don't...can't understand what the
hell it is they want!"

Dawn's face was buried in his chest. She spoke without
looking to him.

"It's our reactions."

"What?"

"It's our reactions they want. They don't have emotions
like us. They're putting us through this to try to understand
how we think and what we feel."

"How do you know that?"

"I don't know, but I do."

Steve looked down to her, trembling in his arms. The
tears began to dry on his face.

"If it happens again, we've got to resist. I don't think
I can make it through much more of this. It's too much;
like they're raping our minds. They'll rape us to death if we
let them."

Dawn collected herself.

"I don't think we can, Steve. They're too strong."

"We've got to try."

"Maybe that's what they want."

"Then, fuck them!" he screamed in exasperation. Then,
shot a glance beyond the illuminated figures to the carnival
of activity beyond where the gremlins remained, more
passive now, and the triangular searcher continued its
methodical canvassing of the desert basin, the gigantic
mothership hovering above. He reined in his emotions.

"Can we make friends of them?"

Dawn shook her head in the negative. "They're neutral; not capable of feeling emotions as we know them. That's what this is all about, Steve. Like the alien abductions you read about in the papers. They want information about us. Who knows what kind or for what reasons. Only this time it isn't a farmer in some corn field in Nebraska. It's us!"

An unconscious shiver passed through Steve as he realized for the first time that he and Dawn were indeed at the mercy of these Beings and that abduction, even physical experimentation on board the craft, was a very real possibility.

"And that thing out there. The triangle," he said uneasily. "What do you suppose it's searching for?"

Dawn forced herself to peek. She spoke, timid.

"Maybe it has something to do with the sound and smell."

He listened to the grinding, boring noise that seemed to emanate from deep inside the earth, beneath the truck and the desert wash around them. The smell, too, seemed to be gaining in intensity so that their nostrils and lungs burned with the pungent, acrid smell of molten sulphur.

"It's possible. I told you about the rare earths mined around here by Alumax and other conglomerates. Could it be they're searching for minerals; mining exotic elements to refuel their craft or bring back with them?"

She recoiled at the sight of the shimmering creatures, then forced herself to study their appearance. Near as tall as she, they stood staring in the camper windows, faces devoid of either lips or teeth with a slit for a mouth and two open holes for a nose punctuating their large, oversized head with skin so white it actually glowed.

But it was their eyes that most frightened her. Large, dark piercing eyes without iris or cornea, that seemed to bulge like an insect's, then narrow into a vortex of chilling emptiness; intense and ubiquitous; totally lacking any semblance of human emotion.

"It could be a way-station," said Dawn, the idea suddenly popping into her mind.

"I don't understand."

"The probe from the mothercraft. Do you see it?"

He nodded, observing the tunnel of light. Perhaps one hundred feet in diameter, it extended from the belly of the huge spacecraft stirring the dust and cactus below like a windless tornado; sucking objects up into the craft even while others were transported down as it scanned the desert basin below.

"Don't ask me why, but I feel like it might have something to do with 'souls'."

"Souls?"

"Yeah. The souls of humans coming and going to and from earth as they are born and die. I read that somewhere. That mystics believe there are special places where the physical and spiritual planes of existence meet. Maybe this place is like that; a kind of way-station where souls pass from one world into the next."

"And what does that make them?" Steve asked, gesturing outside the camper. "Gods?"

"Something like 'gods', I suppose. Messengers or servants from Heaven or, or another place we can't even imagine."

Steve ruminated. He cast a lethal stare outside; his eyes rivetted upon the creatures.

"These are no 'gods'," he spat out contemptuously, "though they'd like us to think they are. No 'god' could ever treat people this way. But aliens, travellers from another galaxy or another time might be curious and cruel enough to put people through this."

"But why us?" Dawn anguished. "Why put us through this? We're nobody special. We've never done anything that could vaguely interest aliens or anything else like that!"

Steve turned to Dawn, the theory taking shape in his mind as he spoke.

"Okay. Let's say that you're right. That these are travellers from someplace else; outer space, Heaven, another dimension; someplace we can't even begin to fathom. But let's also suppose that I'm right and they miscalculated on a voyage between here and wherever they come from and ran out of fuel. So they come here to the Mojave to dredge up whatever minerals they need to continue, but stumble upon us."

"Their first reaction might be to leave, then return once we've left; or even to kill us, if their situation is desperate. But then they recognize me from an encounter years earlier and figure, what the hell? We've got to do this mining operation anyway, so let's see what's happened since. How he's developed. What his mate is like. Exactly what makes these humans tick. Sort of a side benefit. Additional research while they're here. That's possible, isn't it?"

Dawn considered his words for a drawn moment, then gasped.

"Steve, there was more. About the 'gods', I mean. I just remembered where I saw it!"

"Where?"

"Have you ever heard of petroglyphs?"

"Sure. Indian paintings. There are hundreds of them

all through the caves around here. Most people say they were painted by the Piutes, five, maybe six hundred years ago."

"Yeah. Well, I read about them in a book called *Mystic Places* I got from the library last summer. It said that the local indians considered this part of the desert sacred; a place where the deities came to show themselves to tribal priests. It talked about petroglyphs all right, but said the most famous wasn't in a cave. It's right here in the California-Arizona desert near Fort Mojave. Scientists call them the 'Mojave Twins'; a huge ground drawing that can only be recognized from planes passing high over the desert."

Dawn paused to try to comprehend the scope of what she was saying.

"And do you know what they looked like?" She pointed to the curtains, saturate with the image of the illuminated figures parading around the camper. "They looked like them, Steve. I swear to God. The 'Mojave Twins' looked like those 'things' out there!"

1:05 a.m.

The air around them crackled with electricity. Some of it external, but much of it from them, inside, like some hidden self deep within was being fed by or was feeding the creatures outside.

Dawn struggled with the notion of what these Beings were; where they came from and all that was going on around them.

"Steve, this can't be *real*."

"What else is it?" he asked non-plussed. "*What else can it be?*"

Dawn was about to respond, but stopped as if suddenly distracted. She felt the telepathic force like a tug at the edge of her conciousness; subtle at first like a fisherman's bob on the water, nibbled far below if only for a fleeting moment. Still, the feeling registered. It was there.

Steve saw it immediately. In her eyes and the way her body was positioned; rigid, if only for an instant.

"What's wrong?"

"It's beginning."

"What?"

"Them. I just felt it...in my mind, but only a little."

"Are you all right?"

"No," she answered, the dread returning. "I'm not all right."

She began to shake violently as Steve held her tightly, shooting glances to the curtains through which the eyes shone.

The illuminated figures stood still now. Dug in. And Dawn was being pulled inexorably away, back to the hellish rollercoaster of moments before.

"Resist," Steve urged.

It is January 18, 1986 and Dawn is in church dressed in a full princess gown. The priest stands before her. Steve and his brother Ron, the best man, nervously watch as she says, "I do"; Steve kissing her even before he is asked.

"I can't!"

She is a little girl quietly listening as her mother reads a passage from the Bible...

"They're playing with you, Dawn. Playing with your mind!"

She is in her senior year of college at an Alpha Sigma Pi sorority party flirting with one of the boys from Pi Chi...

"I can't! They won't let me; won't let me go!"

She is seven years old sitting in the driver's seat, ringing the bell on her Dad's fire engine...

"You've got to concentrate, Dawn! You've got to try!"

Dawn is dancing with Kurt Hoyas at her high school senior prom...She is emphatic as she makes a point during a high school debating competition...She is singing Christmas carols around the tree with her brother Glen and his newborn baby...She is running the four-forty relay on the track team at Upland High...

"Steve! They're taking me from you! Hold me, Steve! Please don't let go!"

Dawn is joyous...Dawn is saddened...Dawn is exhilerated and angry...and seething with passion...

Then, all emotions drained. Suddenly. As if a plug had been pulled somewhere and the feelings were leaving, replaced by something else.

It is an ugly, hideous thing. The sensation. That memory which she never talked about. Only in dreams did it emerge. A sordid plague of a feeling. But it wasn't a memory. It was all ghoulishly real. Again, she agonized, spinning mind and soul into the deep and ominous grottoes of her darkest childhood remembrance.

Dawn loves to skip rope. Especially in the schoolyard next to her house.

And she's good at it. Best in the whole third grade. Even Laura, her younger sister, says so. 'Course Laura being only seven and a whole year younger thinks everything Dawn does is great. It's nice to have a younger sister, Dawn is thinking as Laura and Meagan sing-out the Teddy Bear.

"Teddy Bear, Teddy Bear, Turn-a-round,
Teddy Bear, Teddy Bear, touch the ground,
Teddy Bear, Teddy Bear, show your shoe,
Teddy Bear, Teddy Bear, that will do!"

Dawn touches the ground singing the words along with them. But still doesn't miss as her eyes divert to the man who shuffles toward them out from behind the school gym. He is skinny with a moustache. His black hair is combed straight back kind of funny 'cause there isn't much of it. He's staring at her. Has she seen him before? She tries to remember. Maybe. But he looks different now dressed in blue jeans and a denim work shirt. She smiles. But he doesn't smile back. Just stares as she skips, loudly singing the words. Showing off. Maybe. A little.

"Teddy Bear, Teddy Bear, go upstairs,
Teddy Bear, Teddy Bear, say your prayers,
Teddy Bear, Teddy Bear, switch off the light,
Teddy Bear, Teddy Bear, say good-night!"

Dawn jumps as high as she can. The man is near her now. Her feet just don't work when she's looking at him. She knew that. It was stupid.

"You missed!" Meagan cries out, the rope tangling between Dawn's legs.

"You never miss! Not that one!" Laura protests.

Dawn untangles herself from the rope. Smiles to the man who watches, embarrassed.

"You're a good rope skipper, ain't you?"

She nods.

"Can I show you somethin'?" He holds a pair of eyeglasses out in his palm for her to see.

All three girls move closer to him. But it's to Dawn that he's talking.

"I lost a lens. Can you imagine the trouble I'll be in if my wife finds out they're broken?"

Dawn thinks a moment.

"That's too bad."

"Oh, it is. It is. They were a gift. And besides," he says squinting, "I can't see too good without them."

Again, Dawn puzzles.

"Where's the last place you remember having it?" chimes little sister, Laura. "That's what my Mom says, 'Think back and that's where you'll find what you're looking for.'"

The stranger glances her way. Smiles a chilling smile. Then looks back to Dawn.

"It was over there. Near those big trees in the back that I lost it." He points. "Will you help me find my lens, Dawn? That's your name, isn't it?"

She nods.

"I know that because I'm a friend of your Dad's. We know each other very well."

She stares. He continues.

"And I know your Dad would want you to help me 'cause I can't see to find them. Not with my glasses broke like this."

She thinks a moment. Looks to her sister Laura.

"Okay," she answers. "I'll help."

The stranger turns to the two others.

"Now, you stay here and play while she comes to help," he says in a firm, angry voice.

They walk.

"Where are we going?" Dawn asks in a small voice.

"Not far. Back there," he says pointing again. "Back to the last place

I had it."

Dawn says nothing more. Just follows. She is confused. Is this man a stranger? She knows she's seen him before. But if he's a stranger...well, Mom and Dad always told her never to be with strangers. But he needs her help. And wasn't she supposed to help people?

They stop. Dawn looks around. She feels a funny, tight sensation in her stomach. The schoolyard is out of sight now. Just trees and rocks and dirt.

He looks down to her, smiling. But it's not a happy smile.

"Do you know I can hurt you?"

"Why would you hurt me?"

"I didn't say I would hurt you. I said I **could** hurt you. Feel." He pinches her hard in the chest. In that spot. Where her breasts are.

Her eyes widen with pain.

"I want to go home."

"No!" he says staring at her, brazen now. "You've been a bad girl and I'm going to have to punish you!"

Dawn stands alone. All alone. Unable to think. Crying.

He approaches. Then, throws her to the ground. She tries to scream, but he covers her mouth, wrestling on top of her. Unbuttoning her shorts, then jerking them down with both hands!

She can't breathe. He's smothering her with his long, boney body. She feels his hot breath on her. His chest is heaving as he grunts sounds like an animal. So terrible. She is burning. On fire. Sick deep in her stomach. She feels his anger. It permeates through her. Why do you hate me? Why do you hate everyone? So much? Then, she remembers. The walk. The moustache and black combed-back hair...

"I know who you are!" she shrieks with her last concious breath.

The stranger stops. Suddenly. He struggles to his feet. Then, looks down at her. Very angry.

"Ten seconds is what you got," he growls, tightening his belt as he speaks. "Ten seconds. And if I catch you, you'll be dead."

Dawn rises slowly. Never taking her eyes from him. She is crying as she pulls her panties and shorts back up. Then, runs. And runs. Out of the woods. Back into the schoolyard in time to see her father, Ed, and two brothers, Glen and Doug, jump up over the six foot fence that separates their yard from the playground. It was Laura, she is thinking from out of the fog of shock and confusion. Her sister, Laura, who went and got them.

The sound of police sirens cut through the hot stillness of the California day as her Dad gathers her in his arms. He holds her close to him. Comforting her. Sheltering his young daughter from the world around her and the horror of what has happened.

"He said he was our friend. Your friend. But he hurt me, Daddy. Oh, Daddy...Oh, Daddy...Oh, Daddy..." she sobs, emptying her soul of all the pain and shame and humiliation...

Dawn leapt from out of Steve's arms, her clothes drenched in sweat; frantic at what she had been through. Anger. Confusion. A sense of violation so profound it left the air around her vibrating.

Speechless, she glared through the curtains, her body immersed in the light as her tormentors continued their soundless procession around the camper. The feelings she was now enduring were identical to those immediately following the rape. And it was numbing. The tumult of emotions set raging through her heart and mind left her psychologically ruined. What had they done to her? What had they taken if not everything she possessed...everything that made her who she was?

"Who are you?" she shrieked out at them. "What kind of 'things' are you? Do you know what you're doing? Do you know what you're doing??" she repeated, the procession stopping to observe her breaking down before them.

It was then that she felt Steve's hand on her shoulder. An odd, wordless gesture, she thought then. Harrowing in its way. So light and yet, at the same time, significant. The touch of a stranger.

She turned slowly. "What, what do you want?" she stammered.

Her husband's eyes were clear, but his expression was fixed and unnatural.

He knelt before her in the center of the camper, an erection bulging in his jeans.

"I want to make love to you."

Dawn moved away from him further toward the windows. He edged yet closer.

"What are you saying?"

He smiled slyly.

"I want you to forget about everything that's happening outside and lay down with me on the cot."

Dawn scrutinized his face in an instant. Like a snapshot the expression registered cold and scary, suddenly transposed with the face of her childhood attacker.

...My God, this can't be happening, her brain was pounding; it can't be happening...can't be...

Dawn's eyes flashed to the outside where the red-eyed gremlins, previously passive, began once again to revel around the truck.

"Steve, this isn't you," she muttered, wiping the tears from her cheeks with the side of her hand. "You can't mean what you're saying."

He paused, rubbing his crotch lecherously. "I want you."

"Stay away," she threatened, the images of her rape clicking away in her mind.

He moved on his knees, sidelong, toward her.

"I need you, Dawn. I need you now."

Outside, the gremlins swarmed, eyes aglow. Agitated like she had never seen them.

"I'm warning you!"

He reached for her. She fended him off.

The thought of her son in mortal danger, of the mental torture she had endured, of the insanity that undermined every thought that came into her head and now this abomination, sparked a pent-up outrage that burst from Dawn like an explosion.

"*You bastard!*" she howled, back against the camper wall, literally fighting for her life as she scratched and clawed to free herself. "*Don't you see what's going on?*"

She clutched the front of his shirt with both hands.

"*Look outside,*" she shouted, pulling the curtains fully open unveiling the smaller beings, faces pressed flat against the windows. "*Do you see them? Do you see the Hell we're living in?*" She shook him violently. "*Our son's life is in danger! He may have been kidnapped; killed for all we know and you have the balls to satisfy your selfish urges!!*"

Dawn's tirade halted abruptly. Her eyes grew wide. She gasped for air, but seemed unable to catch her breath.

"Dawn?"

Steve's expression softened. He looked to her, himself horrified at what was happening.

"Are you okay?"

Her hands fell to her sides. He reached for her. Dawn clutched her heart, collapsing back against the camper wall.

"Breathe. Please breathe," he prayed as the chilling mist began rolling slowly into the camper.

Her face was sheet white. Her body limp.

"I didn't mean what I said, Dawn. It was wrong of me. I knew what you'd gone through and I said it anyway. But it wasn't me! Do you understand that? It wasn't me that thought or said any of those things!"

The dense mist emanating from the illuminated figures outside the camper sent tingles through each of them.

It soothed Dawn, calmed her so that within moments she was breathing normally, the color returning to her face.

2:15 a.m.

Steve cradled Dawn's head in his lap, wracked by a sense of worthlessness and humiliation that left him physically ill. How could he have done this to her? What was wrong with him? Could he really be that callous?

"Please forgive me," he begged. "I swear, I'd never do anything to hurt you."

She closed her eyes.

"I don't know. I don't understand anything anymore," she answered coldly.

It was then that Steve felt the tug like a steel hook sunk deep in his brain, pulling him toward oblivion.

"Steve?"

His eyes raised to the windows alight with the glow of the illuminated figures standing stationary once again around the camper. Sullenly, he stared into their incandescent faces with dark eyes as cold as reptiles'.

"I don't want to go," Steve rasped holding his head now throbbing with the sensation. "I won't! I can't stand it anymore!"

Steve is a boy of seven touring the Goldstone Tracking station where his Dad is an electrician...

Dawn sat up; every fiber in her body revolting at the unfairness of what had happened to her and what was happening to Steve.

"What are you doing? Can't you see what you're doing to him?"

He is at Barstow High in a gym sizing-up a wrestling opponent, the crowd around him screaming; exhorting him on as he lunges forward...

"It's not happening. I won't let it!" He reached for Dawn's hand. "Hold my hand! Hold it tight so they can't take me!"

Steve is in Dawn's hospital room, their newborn daughter, Beth, in his arms...He is in a classroom at American University as a Poly Sci exchange student listening as Henry Kissinger lectures on the new world order...

"Leave him alone!" Dawn screamed, watching as the eyes of the illuminated figures bore down on him and the gremlins clamoured, reawakened by her horror. "You're killing him! Can't you see that? You're killing us both!"

Steve is at a banquet accepting an award for being named "Salesman of the Year" in 1988...He is standing nervously in front of Dawn's dorm room about to introduce himself...He is picking himself up from the ground about to reenter the Cal Lutheran - U of R playoffs after suffering a mild concussion...

"Please, Dawn! Please!"

Steve is five years old and dressed up like Superman for Halloween...He is drinking with friends...He is newborn...He is an adolescent being disciplined by his father...

She reached out, wrapping her arms tightly around him.
"I'll hold you!" she promised. "I'll take care of you!"
But it was no use, Dawn knew, as she felt his huge frame turn rigid, then convulse with terror.

Devastated and alone, she held Steve's body, suddenly lax in her arms. His eyes were vacant.

He was in another place.

Steve's lungs fill with the crisp morning air. Above him the sun beams its warmth down upon him. His eyes scan the blanket of black timber that surrounds him; a streak of yellow aspens running through it like a vein of ore. He is in the Northwest country, he realizes immediately; Oregon or maybe Washington state.

He asks himself, 'Why am I here?' and 'How did I get here?' drawing an immediate blank. Then tries to convince himself that the answers will come. When? Soon, his mind retorts, suppressing the jagged edge of panic that lives just beneath the surface.

Point is, he feels at home here; remarkably at ease. And hungry. Voracious, in fact.

Now he must find answers. He walks along the grassy ridge and into the hardwood-conifer forest sensing something he recognizes as the smell of burning wood. A campfire. He follows the scent, plodding through the damp earth and underbrush until the thick netting of treetops finally gives way to sunlight.

He stands at the edge of a clearing, his large muscular frame looming as he scouts the area acutely aware of, what? Danger? Perhaps, his instincts tell him. He sniffs at the smokey air, the aroma of rabbit skewered over burning mesquite clings to it deliciously.

There is no one in sight. The firepit crackling, encircled by rocks, smolders unattended. The clearing seems deserted. Not a soul in sight; not a tent or sleeping bag to be found.

He enters, ever cautious, angling toward the fire. His hunger is a throb in his gut; a nagging pang that must be sated, or hunger will give way to weakness and weakness to…No, he is a survivor, he tells himself, and a fellow woodsman would never deny a man in need.

His footsteps become more certain. But again the jagged edge rises up from his subconcious. Tracks. There are none. The ground surrounding the fire is smooth with not a twig or leaf disturbed; not even a footprint. Why?

Again, the feeling of unease. And again, the retort, does this change anything? Food is essential. No matter where you are or why you are here. The need to nourish is more compelling than any argument.

Snatching the skewered rabbit from the firepit, he rips loose a piece with his teeth. The gratificaiton is immense. He savors the juicy mouthful. So hungry…so fucking hungry, he is thinking assuaging the low dull throb that lives in his stomach; knowing somehow that the hunger will always be there. But for now he is content and full.

He pauses. Sounds. Campers? Steve wonders. He discards the carcass standing on the balls of his feet, straining to hear.

A welter of concern passes over him as he runs from out of the clearing, through the apron of forest and back to the ridge. He looks down into the sprawling meadow. Then, he hears it. The sound of a rifle shot. And in that instant he remembers how bears are baited with a fifty-five gallon drum filled with beaver meat. Baited, then tracked. Tracked, then shot, he realizes as he feels the sting of a 180-grain bullet in his left leg!

It isn't a direct hit, but the force of the bullet spins him halfway around. The shock sends waves of impact radiating through his body. It disrupts his entire nervous system so that him arms go numb and both legs fall out from under him.

Instinctively, Steve rises from the ground. A flood of questions roar through his conciousness. But only one question and one answer matters. 'Why am I here?' His mind resounds. And the reply, so simple. So direct. 'You are here to be hunted.'

The awareness sends him flying in a blind panic. He lurches from the ridge streaking back into the forest. He had to move and keep moving, he knew, for the hunters would be after him. Assuming he'd been wounded; waiting to put him out of his misery.

He stops to examine his wound too shocked to feel pain. Rather, the sensation is dizzying. Perhaps from fear. As likely from a loss of blood because his left leg, pierced through the calf, is bleeding horribly.

He rips loose a section of shirt to form a tourniquet. Still, he can't stop. Not now. Fleeing is a matter of survival, pure and simple. But where? In what direction? He looks to the sky, overwhelmed. Into the forest. The deeper, the better. And downward, east, into the sun, his shattered mind races.

Frantic, he runs. Long and hard, until he hears a sound that causes his pounding heart to sink. No longer was there a chance to hole-up in concealment. It was the sound of a death knell: Hounds baying in the distance!

Steve runs wildly through the dark wet timber over forest shrubs and through the tangle of vines and low hanging branches; his mind operating at levels of perception he never knew existed. He is hot and flushed and his burning flesh feels liquid, oozing his physical nature away, freeing some instinctual creature buried for milennia.

There is something nearby, his mind tells him. An important something telegraphed by sound and smell and feeling. A running creek.

He veers toward it; pushing his body; commanding his wounded leg to obey. It is life or death, his brain is shrieking as he nears it. Thirty feet across and no less than two feet deep, he is thinking. His chance to throw the dogs off his scent.

He collects himself, then bolts across the creek. He takes several steps in the opposite bank, then doubles back to its center where he treks upstream some twenty yards, then crosses full way.

The barking of the lead dog and baying of the others is more feverish now. They are on to him. Behind him less than one hundred yards leaving him mere seconds to ghost away in a tree or the dogs will rout him out, then tear him to pieces.

He catches sight of a fan-shaped oak tree. Gasping for air, he reaches up to its lowest branch, then pulls himself onto it. His torn calf is beginning to ache now and he grimaces as it bends. He takes another branch into his right hand, then does the same. His climb comes to an abrupt halt when he spots the lead dog. The hound has arrived at the bank of the creek and is sniffing.

Steve wants to watch but forces himself to turn away. The sound of the barking dogs elevates noticably as three others arrive. They sniff the area, noses to the ground; their frustration apparent. They have lost the scent.

"He's a smart one, ain't he?" says the first of the hunters to arrive at the bank.

The shooter, who holds a torn remnant of Steve's shirt walks over to the lead hound, waving it at his nose.

"Now, we'll just see about that, won't we?" he answers, reserving judgment as the dog's tail stiffens.

Steve steals a peek in their direction in time to see the lead hound darting toward the tree where he hides. Within seconds, the dog stands on its hind legs, joined by the others, front paws up on the trunk, heads thrown back barking and growling their fervor for the hunt.

The shooter turns on his heel aiming the .300 Winchester at Steve's heart, then fires.

Steve takes the bullet in his right shoulder. It shatters the bone hitting at such high velocity that it jellies the flesh surrounding the wound channel.

He falls from the tree. The hounds rush upon him, barking wildly. He is totally without feeling now. Though his mind still operates, it is with a distancing that makes the pain and tearing of flesh seem to be happening to someone else. He thinks of Dawn and how lovely she is. And little Steven and Bethany, their baby, and it fills him with profound sadness to think he will never see them again.

When the hunters arrive, the dogs are restrained. The shooter pulls him up by the hair.

"Good size, ain't he?"

"Yeah," the series of rumbled voices comes back from the others.

He feels the jab and puncture of a cold steel blade just below his sternum. Then, hears the tear of his flesh as the knife rips downward to his pelvic bone and a gush of warm blood floods his thighs and genitals.

The last thought that crosses Steve's mind comes quietly and with a vague sense of surprise.

My God, he thinks, they're skinning me alive.

Steve groaned, groggy and disoriented as if coming from out of the deepest of reveries. He was alone...He is dead...disoriented...in a camper in the Mojave.

He ran a trembling hand over his sweaty face attempting to get his bearings. It was a dream. All of it. A terrible, horrible nightmare. But now it was over, he reasoned, eyes racing to Dawn beside him, then to the camper windows, unleashing a howl of agony and despair that rang through the valley. For staring in at him were the ubiquitous Eyes; hungering for information; tearing, ripping his mind apart.

THREE

Mojave Campsite
October 22, 1989
3:00 a.m.

"It's okay, baby. It's okay," Dawn comforted.

"No! No, it isn't!" Steve protested shaking his head from side to side; trying to dispel even the vaguest remnant of his vision. "I was hunted. Like an animal. Then..."

"It's over now. It's over now."

He looked into Dawn's face hungering for reassurance; his ravaged mind desperate for any semblance of sanity.

"They mutilated...my body..."

But such succor Dawn could not offer for she, herself, was now deadened with shock and exhaustion.

"You..."

Her face was expressionless.

"It was rape. I told you before we were married. When I was a little girl, just eight years old, a man. Someone from town took me behind the schoolyard near where we lived and raped me. And now, I've been forced to relive it. Not to remember. But to go through it all again!" Her voice quavered. "Why are they doing this to us, Steve?"

They held one another. He kissed the tears from her cheeks.

"I don't know why. I only know that I love you, Dawn. I truly do."

"I know. And I forgive you for what happened before. It wasn't you. It was them. They made you do it. The way they made me relive the rape and the torture of my child. And I hate them, Steve. I hate them for what they're doing!"

Steve glowered out at the illuminated figures and beyond noticing now for the first time since the onset of the invasion that the smaller ones, the gremlins, were beginning to recede.

He pulled away from Dawn to see more clearly; to make certain it was real. But without a doubt, from behind the shimmering wall of light cast through the windows by the illuminated figures, one at a time, and then in packs, they scrambled along the desert wash back into the valley from where they came.

"What? What is it?" Dawn asked.

"They're leaving!"

"What?"

"Look!" he exclaimed pointing, then tracking their path with his finger. "They're going back into the valley!"

Dawn watched, a feeling of immense relief welling up inside her.

"It's ending, isn't it?"

A smile crossed Steve's lips.

"Yes. I think it might be..."

Together, they observed the exodus as the gremlins scurried back into the night dozens at a time, their grayish-blue bodies hunched over as they receded, heads turning back toward the camper only occasionally with a flash of their laser-like, red eyes.

"We've done it! We've outlasted them!" Steve proclaimed as they hugged one another.

"I know! I know how it will be!" Dawn affirmed. "The little ones will go first because they were the first to land. Then, the monitors. Then, the tall ones. And finally, the ship. The space ship will leave and take them all away!"

"God! God help us! I think you're right, Dawn. And the time..." He looked to his Timex Sportsman. "Jesus, it's only 3:00 a.m. But it's got to be daylight soon. They'll never stay beyond that. How could they? Someone else is bound to see!"

Together, they huddled, joyous in the back of the camper shell watching as the illuminated figures, too, began backing slowly away from the truck.

"You know, if what you said before is true—that they didn't come for us, but were here to do something, some kind of mission, but stumbled onto us by coincidence. Couldn't it be that they've done whatever testing, psychological testing, they wanted and now they're through? Because look, they're going deeper now; deeper into the desert and farther from us. And if you look close, you can see the objects they're beaming more closely than ever before."

Dawn showed Steve the circular probe transporting defined images now; outlines of animals; dogs and cows and donkeys; up into the craft as the inverted triangle of lights worked in tandem with it, intensifying its seemingly inexhaustible search of the desert basin. The light that laced the underbellies of the six smaller craft attached to the mothership appeared also to be part of the operation, flashing signals in coordinated patterns; perhaps even orchestrating the arcane activities that continued widespread and unabated.

"Maybe their mission here, whatever it is, is ending, Steve. Maybe they'll just leave us. Alone. Unharmed. Isn't that possible?"

"Yes. Yes, it is possible! What could they want with us, really? Who are we? I mean, if they wanted to take human specimens, they sure as hell wouldn't come to the middle of the Mojave to find them."

Dawn smiled a rapturous smile as a deep and pervasive

euphoria seeped into their hearts and she began praying.

"Thank you, Lord, for allowing us to be spared. We believe in you. We trust in your almighty wisdom. We are your servants and pray now that we will be united once again with our families and children, who we love more than anything on this earth."

Steve's emotional state had risen to the point of intoxication as he cheered the disappearance of the gremlins and near withdrawal of the others.

"This is it!" he called out, his happiness unrestrained. "They're leaving! *We're going to live! We're going to see our children again!!*"

The cries of joy and prayers of thanks reverberated within the tiny camper shell. And still, for an instant, after they stopped so suddenly, the muted echo remained. Then, silence. Total and complete. For it was happening again. But this time with a vengeance. As if these pathetic moments of ecstasy had been created simply to be snuffed immediately after they had been supplanted.

The illuminated figures stepped forward once again, their probing stares more absorbing than ever. And the gremlins, hundreds of them, charged forward surrounding the camper; overrunning it as they had never attempted before.

Steve and Dawn watched what was going on around them in disbelief. The intricate world of hope and dreams for the future that had been created in their minds was being snatched from them now before their very eyes. Hopes of seeing their children and parents and friends; expectations for their own lives and the beliefs they cherished were being laid to waste. In an instant, joy became grief. Hope became despair. And the gratitude they had voiced reverted to rage.

Steve snapped, a murderous fury spewing from out of him as he pounded the camper windows with closed fists.

"I'll kill you! I'll fucking kill all of you! Every one of you!" he bellowed.

"They'll never let us leave here," Dawn muttered in a lilting voice. "Never, ever let us leave here alive. They're going to kill us. What choice do they have? We've seen them," she babbled. "We've seen them so they can't let us go or we'll tell; we'll tell everyone. We'll tell the world who they are and what they've done."

Steve reached for his twelve gauge.

"You sons of bitches!" he cursed, taking the shotgun in his hands. "You think you can get away with this!"

But before he could raise the weapon, the monitors glided forward and the Voice inside of his head began, drawing his sense of purpose from him; arresting his will so that the fury subsided and he became powerless to hurt them, 'Don't do anything to hurt us.' Then, 'Put the gun down. You have no chance.' And finally, measured and reserved, 'In the end, we'll kill you if you try to harm us.'

Steve dropped the shotgun. It made a clanging metalic sound as it stuck the camper's steel truckbed. It was no use resisting. No use at all, he realized, his feelings of rage replaced by a sense of saturate horror as he looked out into the valley where the tunnel of light shooting down from the mothercraft, tore across the desert expanse on its way to their camper!

3:25 a.m.

The gremlins, droves of them, clamoured, streaming over the top of the truck pressing their grotesque faces expressionless as leather masks against the windows to taunt their captives. The procession of illuminated figures continued unrelenting as dozens of others joined them; some standing stationary just five to six feet from the truck while others observed like spectators from afar. Yet, beyond the frenzy, a storm was going on within the tunnel of light as Steve and Dawn watched the probe approaching, dumbfounded.

There were properties the probe possessed unlike any he had ever imagined, Steve noticed now for the first time since he'd observed it. Following in the wake of the "searcher", which he judged to be a kind of homing device to identify and locate, the probe cut a swath of radiance through the night, but it was not a blinding light, he realized staring into it, eyes unshielded and apparently unaffected. And now he could see that it was not simply objects that the probe was retrieving, but elements; fine showers of particles being drawn up from the ground.

The grinding sound intensified and the tunnel of light they had observed earlier from a distance was now actually causing the truck to vibrate as it neared. It was a laser of some kind, Steve believed, doing massive drilling at some extreme level deep within the earth. But more, he realized, fear rising up like a knot in his throat, this would be the mechanism the aliens used to take them on board the craft!

"They're going to take us," Dawn croaked, barely able to speak.

"I know."

"The children...the poor children. I'm so sad."

"They'll understand."

"How? How will they understand when their parents have simply vanished from the face of the earth!"

She was crying.

"I have this feeling. This, this dread of being captured

and torutured. Steve," she confessed, "I think I'd rather die here in this camper than go up with them."

They watched the probe edging forward in short, paced increments. The truck was vibrating all over now. Shaking violently as the rank odor of burning metal infused the camper shell.

"I don't want to die, Dawn. I've never been more scared. What can we do?"

"Suicide is a sin. But God would understand. I know He would because I just can't bear the thought of those things out there touching me and probing my insides. I've been raped once, Steve. It will never happen again!"

"I'll stand by you. I'll be strong, Dawn. I promise."

They each held onto the sides of the camper shell now to keep from being pummelled. Camping gear and silverware and canned goods cascaded from out of the lowboy. The truck was actually being lifted; lifted off the ground!

"It's pretty near on top of us now, Dawn. What's happening? What the hell will happen when we're inside the tunnel?"

"Help! Help! Someone please help us!!" Dawn was screaming hysterically as the truck began rocking so violently it seemed about to explode.

The noise. The grinding sound so deep and reverberating with such pressure was maddening. The pungent stench, thick and acrid in the trapped air around them seared their lungs and stung their eyes. And now the gremlins totally wanton; agitated and agressive as they stared in the windows with unbridled glee; clinging to the top of the camper shell; attempting to overrun the monitors from behind the tailgate. And all the while, the illuminated figures stood, stationary once more, observing; studying the hysteria both inside and outside of the camper.

"Get the fuck away from us!" Steve screeched at the gremlins, through the hellish sounds and sights that surrounded them. "Leave us alone! For God's sake," he pled, "leave us in peace!"

He clutched the side of the bolted down bunk leaning toward the left window. His back was to Dawn who watched aghast as one of the illuminated figures stepped through the morass, beyond the gremlins, raising a distended, three-fingered hand in her direction.

Dawn was too wooden to react, but with that motion a fleck of light the size of a cigarette butt passed from its fingertip through the window and into her abdomen.

"Ahhhhh!" she shrieked at the sight of it.

"What is it?" Steve called out over the bedlam.

"My God, it's inside my stomach!"

She tore away her shirt with her free hand frantically trying to brush it away. But it was not on her. It was inside her body.

"We're going up!" Steve bellowed, the probe's brilliance totally enveloping the truck and the area around it.

"Jesus, my God and Savior, help your humble servant. Save us from all danger. Protect us from evil!"

Steve clenched his teeth certain they were about to disintegrate; reduce to the atoms and molecules that comprised them once inside the tunnel.

Dawn speculated that she'd been impregnated, or that an ovary had been plucked from inside her, and they would be taken, intact, for experimentation.

But the reality was different from anything either could have imagined for now both believed the truck and the entire environment around it was being lifted up into a vast collection center inside the mothership. Above them, they could see the underbelly of the craft camouflaged by clouds as they were being lifted; its immense banks of lights continually blinking and then; nothing!

"Do you feel it?" Steve hollered over the dissonance. "It's like the earth has been cut out around us. All of it. The truck, the sky, the desert wash; everything!

"There's air beneath us. I feel it, too. The scenery hasn't changed, but it's like we're suspended somewhere. Like they've separated us from earth and put us into some kind of vast museum."

Steve shivered, wrapping his arms around his upper torso.

"And colder. Do you feel it?" he asked. "Suddenly now. Like oxygen is, is pouring in from somewhere," he stammered peering outside through the windows in wonderment, not knowing what to expect next.

Strangely, despite their ascent skyward, their environment had indeed remained the same. They took an accounting: the desert, the mountains, the camper and surrounding brush. But there was no noise. The drilling had stopped and the smell was gone. The spacecraft, so ominous throughout their captivity, was no longer above them. And the gremlins truly were retracting now, their dwarfish bodies disappearing one by one; beamed up as lights just as they had come. And the illuminated figures, too, had all but vanished. The virulence and mental anguish they'd inflicted ushered out by a new totally translucent presence, unmistakable in its femininity.

Now, clearly, there seemed an 'order' to this realm of existence that had so suddenly and pervasively enveloped them: the gremlins so crude and frenetic; the short, stump-like grays who monitored them; the more intelligent, scientific illuminated figures and now 'she,' this presence so gentle and yet so powerful in nature.

It was this entity who had dispelled the lesser creatures, Steve and Dawn understood immediately, watching in astonishment as the smoke-white, swirling form descended from the left side of their sky. A Being that in other times may have been called an 'angel'; she was radiant; exquisite in appearance and nature, her lucid robe flowing as she floated gracefully toward them.

And then, a voice of another kind, soothing; comforting and totally serene.

'It's all right. I'm here now to protect you. Be at peace. It's almost over.'

With that the presence left as she had come. Mysteriously lifting up and away from them. Still watching over her two charges and minding the lesser life forms as she faded from view; making certain even to the end that they were safe and comforted and their torment was over.

"We've been chosen for something," Dawn confided to her husband. "I don't know what. Maybe it's just to tell our story. What we've seen. What we have been allowed to see."

Steve looked out into the expanse of desert suddenly tranquil. The 'searcher' was the only alien presence remaining, except for the two monitors that stayed positioned beyond the tailgate and the illuminated figures, who had backed off from the truck and were retreating back into the desert.

"I still can't believe this has happened," he uttered absently. "It seems like a dream, but I know we're not dreaming."

With the moment's respite came a feeling of total mental and physical exhaustion. All telepathic communication had ceased and in the absence of the freezing fog that alleviated bodily discomfort both became aware of an excruciating need to relieve themselves. Still too terrified to leave the camper, they urinated one at a time in a plastic container.

"Is it all right to sleep?" Dawn asked afterward.

"I can't stay awake any longer," Steve answered wearily.

"But they're still here. The monitors. The 'searcher'."

Steve, too fatigued even to speak, simply layed down, clothes and boots still on, atop the bed.

Dawn glanced from the camper window into the western sky one final time before laying down to join him.

Far above them, a bright light bled through the black firmament for the first time since the beginning of the onslaught. As they had come, the demonic, red-eyed creatures were being beamed up from the outer fringes of the valley to that one point, until it started resembling the moon; a crescent moon that began swinging like a pendulum.

"They've given us our moon back," Dawn warbled to no one.

This pleasant thought the last she would remember before lapsing into a deep and dreamless slumber.

8:00 a.m.

The desert sun splashed into the camper shell from the windows and truck's open tailgate. Somewhere the distant sound of a single engine prop plane could be discerned as it passed overhead; the first conventional sound the Hesses had heard in more than twelve hours.

Steve stirred, the remembrances of the night before so vivid that even before his eyes opened, he bucked up in the bed as if awakening from a nightmare. Tired and feeling mentally blunted, he doused his face with bottled water.

Dawn awakened a few moments later feeling that same mixture of lassitude and anxiety. She sat up in bed, groggy. Her body lunged forward as she gasped, then caught herself at the morning's first recollection of what had transpired only hours before.

Her panicked eyes shot to the windows reflexively.

"They're gone," Steve offered. "No sign of them anywhere. The valley's empty. Not a cloud in the sky."

Dawn took a gradual account of herself and their situation.

"Thank God," she muttered.

Still dressed in the jeans and workshirt he'd gone to sleep in, Steve left the camper to search the desert and surrounding wash for any physical evidence that may have been left behind.

Dawn shook herself from out of her stupor, then joined him, hopping out of the camper and into the cab where she turned on the radio, then fished across the AM and FM bands for news of what had happened.

"What day is it?" she called out to Steve.

He glanced first to his wrist, but his watch had stopped; then up to the sun.

"It's 8:00 a.m., or near to it," he answered on hands and knees as he scoured the area around the mesquite tree. "And it must be one day later; the morning after they left us."

"That's not what I mean," she yowled back, frustrated at her inability to find news of the event. "I mean, how do we know what day it is? Or what year; or decade for that matter. Everything's been turned around. For all we know, World War III might have happened and we don't even know it!"

He approached the cab, wiping his hands clean on the sides of his jeans.

"Nothing. Not a thing. Camper's untouched. So far as I can tell, not a twig around it's been broken; not a grain of sand moved. Nothing!" He looked to her, vexed. "How about the radio?"

Dawn turned the volume up on WFLG, a local country station, where Randy Travis crooned "Diggin' Up Bones".

"Hear for yourself," she said disgustedly. "No mention of anything even vaguely related to what happened. Not even on the news channels!"

Steve grimaced.

"Move over."

"What?"

"I said, 'Move over,'" he repeated.

She did, watching as he jumped into the driver's seat, then turned over the ignition.

"Where are we going?" she asked leerily.

He took the truck up the rugged mountainside leading out of the valley.

"To find witnesses," he answered, racing the Ford pick up onto one of the unnamed trails they'd travelled the night before.

They drove six miles on Black Canyon Wash before coming upon another campsite. Steve parked hurriedly, then rushed from out of the truck, slamming the door behind. Dawn followed.

"Now we're going to find out what they hell is going on around here," he abraded.

They approached the first camper, a man who looked to be in his sixties, crouching over a campfire as he brewed a pot of coffee.

Steve was about to ask what he'd seen in the sky the night before when suddenly the folly of the request struck him. What if he'd seen nothing! Judging by their appearance and the strangeness of his request, he'd be lucky if the man didn't contact the police once they'd left.

His gait slowed. His self-assured bearing waivered as he reassessed his strategy.

"Morning," Steve greeted in a quiet, cautious voice.

The man looked up from the campfire. "What can I do for you?"

Steve and Dawn moved nearer.

"Were you camping here last night?" Steve casually inquired.

"Yep," the camper rejoined. "Why do you ask?"

Steve stopped a few feet from him, Dawn at his side.

"Did you notice anything peculiar. In the sky, I mean. Anything abnormal?"

The gray-haired man ruminated for a moment, then shook his head.

"Nah. But that wouldn't be so unusual. Not if it happened after 7:30 'cause that's what time the wife and me go to bed." He turned, then hollered over his shoulder. "Ain't that right, Bobbi?"

"What's that?" They heard a voice come back from behind the canvas tent flap.

Then, the second camper, a lively looking senior with her hair pulled back and into a bun, poked her head from out of the lean-to.

"Got a fella and his ladyfriend here want to know if we saw anything unusual in the sky around here last night," the man explained. "Told them we turned in early."

"That's right," she affirmed. "See, we come here for huntin'. Early to bed, early to rise. We leave the stargazin' to young folks like you."

The man chuckled from beside the campfire at his wife's forwardness as Steve and Dawn turned to leave.

"Want some coffee?" the woman called out after them. "You two look like you could use some."

"No. No, thanks," Steve replied just before he and Dawn got back into the truck.

Steve started the engine.

"They didn't see a thing," he grumbled. "Far from everybody and in the valley like we were, it wouldn't surprise me if no one saw anything!"

"Steven!" Dawn exclaimed, the nightmare remembrance of her young son's torture suddenly surfacing from her subconcious.

There was no further explanation required. Steve knew exactly what she meant.

"You're right," he agreed, a sense of urgency coursing through him. "Let's get to a pay phone and make sure he's all right!"

Steve took the truck north up Black Canyon Wash back toward Wild Horse Canyon Road. His foot pressed down hard against the gas pedal, trying to make time when suddenly Dawn felt him slam on the brakes.

Dawn's lithe frame whipped forward toward the windshield, then back into the passenger seat. She turned on Steve about to complain, then stopped abruptly.

Steve's hands were clamped around the steering wheel. His entire body was quaking as he sat rigid in his seat, a cold sweat oozing down the sides of his face.

"Steve? What is it?"

She looked to him, then over his shoulder beyond the roadside where a triangle of lights hovered low above the mountaintop.

They had escaped the valley near Tabletop Mountain, the torturous realization came to her, but the 'searcher' had followed!

"Can you drive?"

He nodded.

"Then, let's get out of here. It can't follow us once we hit the highway," she feverishly reasoned. "It can't follow us into L.A.!"

Steve took a deep breath. He composed himself, then put the truck in gear, feeling once again the tingling, chilling sensation of the cold mist that had enveloped them earlier that morning and during the most devastating periods of the night before.

"Do you feel it?" he asked Dawn nervously.

"It started in my legs," she confirmed, stomping her feet on the floor of the car, "but now it's running all through my body!" She looked to him, frantic. "Why is that, Steve? Why would they do something like that?"

Steve didn't respond, but gunned the engine bounding the truck east on Blackhorse Canyon Road, the 'searcher' still trailing along side them. But for what reason? he wondered, Dawn's question caroming wildly within the depths of his soul. To intimidate? To locate? To gather information? The two of them could only speculate, watching and praying for the nightmare to be over.

• "Please! Don't go so fast!" Dawn cautioned. "We're going to crash!"

"We need to get to the interstate," he shot back impatiently. "Who knows what the fuck that thing can still do to us!"

She shivered.

"What about Steven? When can we stop? I'm so frightened for him!"

"It's got to leave us once we get into a more populated area. It *can't do anything then!*"

Dawn's frenzied mind wondered what the aliens could and couldn't do, warily watching from the passenger window as the object followed along side of them beyond Blackhorse Canyon Road and onto I-15 where at last she felt she could breathe again: The 'searcher' had disappeared!

Steve stopped at the next gas station where Dawn called home.

"Good news," she reported upon returning to the car. "They said it's been an uneventful weekend."

Steve started the truck, senses too deadened even to comprehend the irony of her words much less react to them.

He pulled back onto the highway, then spoke abruptly in a choppy, mechanical cadence.

"We can't tell Wolfy and Diane. We can't tell anyone. They'd never understand. They just wouldn't."

"Well, it seems like we need to tell someone," she objected. "We've got to report it."

"To whom?"

"The Air Force. Or the FBI. I don't know."

"Okay," he mitigated, eyes locked on the road, "let's think about it. But for now, until we get our own heads together. It's better to tell no one."

Dawn nodded her agreement running the fingers of her left hand over the upper part of her neck, near the jugular.

She touched the area still sensitive where two tiny puncture wounds had appeared during the night, then turned to Steve.

"It's not over," she said without emotion.

III: WATCHERS

Do you know what its like to want to protect your family every day and every night and know that you can't even protect yourself?

'Cause whether you see them or not, you can feel them and it's like they live inside you. Inside your brain and they can do whatever they want, when they want and there's not a damn thing you can do about it.

Steve Hess

ONE

La Mirada, California
October 22, 1989
4:35 p.m.

When Steve and Dawn returned to their La Mirada home the normalcy was jarring. Little Steven had spied them coming down Clear Spring so that he and Wolfy and Diane, with Bethany in her arms, were all waiting when they pulled into the driveway.

"Didn't expect you back so soon," Wolfy called to his son, who was getting out of the truck. "And I don't see no mule buck, either."

"Nah. No luck," he conceded. "Not this time, anyway."

Little Steven ran to his mom. She embraced him like she'd never let go.

"I missed you so much," she gushed, impassioned.

"I missed you, too, Mommy."

"And how was Bethy?" Dawn asked hopefully.

"An angel. But you," said her mother-in-law looking frightened. "You look...have you been ill?"

Dawn let go of Steven, then turned away.

"No. Just a little tired," she fibbed, walking along side Wolfy as they entered the house.

"It's my fault, really," Steve interjected. "Mid-Hills was sold out, so we had to sleep roadside the first night, then set up camp near Tabletop the next."

"Sold out?" Wolfy repeated incredulously. "This time of year? That's hard to believe."

"Tell me about it," muttered Steve.

"Well, did you at least make it into Nevada?"

Dawn, who was still fawning over Little Steve, smiled miserably.

"Yeah. I played the slots; and Steve even won seventy dollars."

Diane, the optimist, sipped from a cup of coffee that had been waiting on the console.

"So, at least it wasn't a total loss."

"No," Dawn wearily agreed. "It was fun. But if you don't mind, I think I'd better take a shower and go lie down."

Diane, who was a trained nurse, walked toward her, placing the palm of her long, thin hand over Dawn's forehead.

"No fever." She turned to her son. "What did you do to this poor girl?"

Steve shrugged.

"Bet they never did go hunting," Wolfy jibed. "Probably spent the whole weekend dancin' and whoopin' it up at Whiskey Pete's!"

Diane shepherded Dawn through the narrow hallway into the master bedroom.

Dawn sat on the edge of the bed, then kicked off her Reeboks.

She smiled weakly. "I don't think I'm going to bother with the shower just now."

Diane's expression was keen with concern.

"You and Steve..."

"No. Nothing like that. Just very, very tired."

Diane watched as her daughter-in-law curled up beneath the covers. Dawn was asleep before she turned to leave.

Realizing that Dawn wasn't feeling well and Steve seemed exhausted, Diane and Wolfy said their goodbyes before lunch, with baby Bethany napping and Steve, Jr. in the playroom watching cartoons.

Steve joined his young son. His 6' 1", 225 pound frame eased comfortably into the recliner. At last, he put his feet up, his mind a muddle of fears and theories and emotions, while Little Steven laughed hysterically at the antics of Bugs Bunny and Elmer Fudd on the TV screen.

The contrast between the warm familiarity of the playroom and the calculated cruelty of all that he and Dawn had gone through in the Mojave could not have been more stark. What was he to do now that the experience was over? Was it possible for a man, who had seen what he had, to put these feelings away in some closet in the recesses of his mind and slip back into everyday existence as if nothing had happened? If they had been "chosen" as the final Being seemed to intimate, what was he to do? What was their mission? If they weren't part of some grander plan and all of it had been a chance event; one that could never repeat itself and that only he and Dawn could know; what kind of freaks had this strange and shocking experience left them?

Steve's eyes fell upon his 2 1/2 year old son, hilarious, watching Elmer Fudd, dressed in formal hunting attire, blasting away with a shotgun as he chased Bugs through a cornfield. The sight of it set the memory of his 'vision' loose like jagged slivers of glass coursing through the vessels of his brain.

He remembered the rawness of his desperation while being hunted and the calculated deliberation of his stalker and wondered if he was not that same victim of prey to the creatures he had encountered. The experience left him questioning everything: about humanity and religion; about himself. What a ridiculous man he was! Steve speculated. How petty everything seemed when compared to the magnitude of Them, the monster-saviors, he had met and now believed had been visiting the East Mojave for generations.

He pondered the future as Little Steven switched channels to Sesame Street, then a Bogart movie and the Newlywed Game. The reality of the workload he'd left behind from the Norwalk State Hospital to the MetroRail construction in Long Beach began seeping into his thoughts; and he wondered if he could do it. If he could bring himself to care about any of that again.

Little Steven switched channels to ESPN where the Celtics played the Knicks, then to John Wayne as a soldier in Fort Apache, holding momentarily on an ancient Twilight Zone rerun on a local station.

"Stop!" he screamed.

Little Steven held the channel staring at his dad oddly as Steve watched, totally absorbed by what he was seeing.

Space aliens had landed in Washington, D.C. offering "to serve man", Steve remembered. Strange, but the play on words, after all of these years, seemed suddenly amusing since the story was about a cookbook used literally to serve man as a food source.

"Dad!" Steve complained.

"Just a little while longer," he asked smiling, then laughing and finally bursting into tears.

Rod Serling's aliens were tall with large craniums and distended arms. Clearly, of superior intelligence, they communicated telepathically encouraging their human victims to eat hearty before feasting on them, themselves.

"If you only knew," he muttered, eyes glistening with frustration. "If you only fucking knew!"

"Dad?"

Steve's attention returned from the TV screen back to Steve, Jr., who had been studying him throughout.

"Why are you crying?"

Steve invited him up onto his lap, unable to control the tears that now rolled down the sides of his face.

"I don't know, Steve," he confessed. "The truth is, I don't know why I'm crying."

12:30 a.m.

Dawn's eyes were closed and her body still as she slept, but her mind journied back to the Mojave and places unnamed where, like Steve, she struggled to comprehend what had occurred and what it could all possibly mean.

Their experiences in the camper played over and over again in her mind. The diabolism of the gremlins; the illuminated figures so cruel; the probe that threatened to destroy them; the "comforter" who offered them succor. What world did they come from, so foreign and strange? Were these the instruments of good, or the gods of evil, Dawn agonized, but there seemed no clear solution, for the answer was both.

True, this was not something either she or Steve had asked for. The sense of helplessness to which they had been subjected was worse than anything she could ever have imagined. Yet, to know that it all truly existed; that Beings such as their "comforter" were *real*. What knowledge, Dawn marvelled; what power to treasure in one's soul! Surely, such insight was valuable and needed to be shared with others. But how? Where did it all fit?

Still, one thing was certain. Dawn had undergone a trauma that left her psychologically shattered. The part of her that reasoned knew this. The part of her that intuited understood it also for all through the night, the images so long forgotten or suppressed rose like fetid cadavers from out of her subconscious. Images of her childhood attacker; of her fears about little Steven's survival at birth; of these alien lifeforms that had inflicted such pain.

But there were other events half-hidden that this night began emerging into Dawn's vocabulary of terror: Dark corridors and lighted ones; a pervasive sense of physical violation and always that feeling; the ubiquitous perception that someone was *watching.*

Dawn sprang up in her bed, heart pounding and eyes wide open.

"*What have you done to me?*" she blurted.

But no one was there. Not even Steve.

Dawn peeked at the clock on the nighttable beside her. It was 3:30 a.m.

She slid her feet into a pair of slippers beneath the bed, then crept down the hallway, careful not to wake the kids.

In the kitchen perusing a set of architectural drawings was her husband wearing the burgundy bathrobe she'd given him for Christmas.

"New hours?" she asked gently.

He laid the drawings down on the kitchen table.

"Can't sleep?"

"Since two."

She walked to the fridge.

"Warm some milk for you?"

"Nah. You know me," he abstained. "Once I'm up this long I might as well make a night of it."

"How 'bout some coffee?"

"Why not?"

Dawn plucked two cups and saucers from the kitchen cabinet, then boiled some water.

"You had a nightmare, didn't you?"

He nodded grimly.

"Want to tell me about it?"

"It's not a nightmare, exactly," he explained. "More like reliving the experience over and over in dreams."

She poured the steaming water into his cup.

"How much do you remember?"

Steve's eyes raised from his cup, surprised.

"All of it. Why?"

"What about when we were asleep?"

He guffawed, "Well, no I wouldn't remember that. I was asleep!"

"But they were still there. The monitors, the 'searcher', the illuminated figures, too."

"Yeah?"

Dawn sat across from him at the table.

"So how do we know other things didn't happen while we were sleeping?"

"I suppose they could have, sure," Steve conceded. "But we were in the camper the whole time."

She took a gulp of coffee. Her expression, cool and composed seconds before, was anything but that by the time she put the cup down.

"But where was the camper?"

"On the ground. Near the wash."

"And the ground?"

"I don't follow."

"My point is, I don't think we ever came down from the museum, Steve!" Her chin quivered. "We were on board the craft the whole time. And that's what my nightmares are about. Not the experience. *The time after the experience when we were sleeping!*"

Steve reached for her hand across the table.

"Dawn...anything's possible. My God, I hardly know who I am anymore let alone the things that may or may not have happened to us out there."

"But you. What about your dreams," she persisted. "The ones you had tonight?"

Steve searched his mind to retrieve them.

"Mostly about the visions; the ones in the camper." His eyes narrowed. "But there's another I can't entirely explain."

"What?"

"It's more of an image then a dream...a feeling like walking through a kind of tunnel or passageway."

"White? Like the light from the probe?"

He nodded speculatively. "Yes."

"I had that same image tonight in my dreams!"

Steve took a sip of coffee, then sighed.

"Isn't what we remember enough? Jesus, Dawn, it's not like we've got to go looking for things to worry about."

"No," she answered, leaning toward him over the table, "but if something happened during that time that could help us understand; help explain what it all means..." She looked to him, her green eyes wet and yearning. "Steve, I swear, I think I'm going crazy. I worry. I worry so much that they're going to take our kids or that they've changed me in some way; or made me pregnant!"

"You mean what happened toward the end, in the camper?"

"That; and other things."

"What?"

"I didn't say anything this morning because I didn't think it was important, but now I don't know."

"What is it?"

"Marks," she answered with lethal seriousness. "Two

red dots. There were scabs on them this morning."

Steve stepped around the table to examine them.

"How could something like that have happened?" Dawn asked as he touched them.

He shrugged.

"An insect bite, probably. Nothing to worry about, anyway."

Dawn gazed down into her coffee cup.

"I'm not so sure," she answered softly.

"What do you think it could be?" he asked, walking back around again.

"I don't know. But it scares me. The thought of them putting something inside my body scares me to death."

Steve downed the last of his coffee. He plunked the cup down in front of him.

"Then, you've got to see a doctor," he concluded. "I mean it. You've got to put something like this to rest once and for all or it will eat at you forever."

She smiled across the table at him.

"I will."

3:45 p.m.

Dawn lay beside baby Bethany on the thick carpet of the playroom floor.

"Does Bethany like rabbits?" she cooed, brushing a stuffed animal over the infant's cheek.

She smiled lovingly.

"Oh, she likes rabbits!" Dawn laughed, touching its soft coat against the skin of her face. "Bethany loves her rabbit!"

This was Dawn's favorite time of day. With Steve at work and Little Steve at her mom's, the house was at last quiet and she could devote time to Bethany; just the two of them.

She rolled on her back with Beth on her stomach.

"Do you want to go for a ride?" she sang to her. "Baby want to go for a ride?"

Dawn snuggled up to the infant, staring deep into her eyes. Then, held her there a prolonged moment.

"What?" she wondered aloud, setting Bethany down at her feet as she stood to make certain she had it right.

With windows and doors shut on a breezeless southern California day, the curtains in the room were fluttering. And then, as she stood there in the center of the room, she felt it; a freezing cold wind running through her long, chestnut hair.

Dawn made an involuntary circle around her baby, glaring as she examined the four corners of the playroom.

"I know you're here," she warned the invisible presence that had entered. "I know because I can feel you here in the room with us. So listen to me now: you can do what you want to me and my husband, but leave our children alone!"

TWO

La Mirada, California
October 25, 1989
6:00 p.m.

Steve came home from work on Wednesday evening to find Dawn where she had been for the past two nights, in the study buried behind a stack of books concerning UFO's, extraterrestrials and spiritualism.

He walked over behind her chair, then pecked her on the cheek.

"Whatcha readin'?"

She held the book cover up in front of him.

"*Intruders.* It's about alien abductions; and there are others," she hastened to add, waving a free hand at the pile. "I've even got my brother, Glenn, downloading information from UFONET."

"UFONET?"

"Yeah," she chortled incredulously. "It's a worldwide computer network!"

He nodded glumly, then shuffled to the door.

"How was work?" she asked looking up from her book for the first time.

"It's this MetroRail," he groaned, undoing his tie. "It's already weeks off schedule." He shook his head. "And you know, after this past weekend, I can't say I really give a damn. The whole project; everything seems so... insignificant. So *pointless.*"

"Well, dinner won't be ready for a while yet. So why don't you wash up; take a shower." She winked. "It'll make you feel better."

That night at the dinner table, Steve's attitude seemed

to have turned one hundred and eighty degrees. He was more sure of himself, even optimistic, as he served up portions of chicken and vegetables to Dawn and Little Steven.

"You know, I've been thinking a lot about what happened this weekend," he began, passing a plate, "but it wasn't until just now that it came to me. A way of approaching all this that might help us both."

Dawn took the plate from him.

"Okay. What's your idea?"

He snared a chicken leg from out of the casserole dish.

"Fact is, I don't think either one of us is faring too well lately. Neither of us sleeps anymore. You're turning into some kind of UFO fanatic and I can't do my job worth a damn." He placed the chicken leg on Little Steven's plate. "That's not us, Dawn. It's not what we're about and it's got to change."

She was mincing the chicken on Steve, Jr.'s plate as she spoke.

"Go on."

"Anyway. I believe you've got to face challenges straight up or you'll never overcome them. So, I'm suggesting we do two things." He raised his fork from the table. "The first is to go back to Tabletop Mountain to see it again; where it happened; and if we're lucky, maybe even why it happened. We've got to come to grips with what happened, Dawn, because one thing's sure, we can't make it go away. It's too much a part of us. We're two different people than who we were because of it."

Steve tried to assess her reaction. There was none.

"And the other? You said there were two."

"The second is that we get this off our chests. We can't hold this inside forever and so my second idea is that we tell Wolfy and Diane and Ed and Bonnie everything that happened last weekend." Having finished serving, Steve turned his attention to his own meal. "Hell, they're our parents, right? If we can't trust them, who can we confide in? And, God knows, Dawn, I feel like we both need someone to talk to about all of this." Steve waited for a response. Finally, he asked, "So what do you think?"

"I think it's dangerous."

"Which one?"

"Both," she retorted. "I don't know how you feel about it, Steve, but I feel like we barely got out of there alive!"

"So, you run from it? Pretend it never happened?"

"No!" Her fair complexion reddened. "I'm a grown woman with two children who depend on me, Steve! What if something happened? It wouldn't be just you or me. Think about them and their future."

"That's what I am thinking about," he insisted. "Don't you think I see what's happening?"

Dawn's temper cooled as she contemplated the effect the experience was already having on her family. How ragged they'd both become.

"And the second idea?" he asked, more calmly. "What about that?"

Her attention turned to the plate on the table. She fidgetted with her potatoes.

"I'm not sure," she confessed. "On the one hand, I'd like to be able to discuss it with someone; especially my parents. On the other, I don't want them to think less of me for it." She fumbled for the words. "What I'm really trying to say is, I'm terrified they'll think we've gone crazy."

"Even if we are crazy. They're still our parents," he said reassuringly. "You'll still be their daughter or daughter-in-law."

"I know," she agreed, taking a deep breath, "but even now with you here at home, Steve, I'm shaking like a leaf just talking about it. No," she said shaking her head, convinced. "Not now. Not yet. I'm just not ready."

"Well, all I'm asking is that you think about it—whatever we do, we need to do it together." He smiled wanly. "Let's face it. We're each other's best friend. Without you there'd be no one to share my feelings with."

"Likewise," Dawn admitted. "Just one of the many fringe benefits of the experience, I guess."

Steve took a gulp of milk.

"By the way, did you get to the doctor like you said you were going to?"

"Yep," she answered, swallowing. "All checked out this morning. Bonafide 100% healthy. Doctor Wallach couldn't really tell what caused the marks, but he said it was nothing to be concerned about."

"And the other?" asked Steve between forkfulls. "The problem with the bleeding?"

"It's not so unusual to spot every so often, especially this soon after Bethany."

"Is that what the doctor told you?"

She nodded.

"So, what causes that? To spot, I mean."

"Nerves," answered Dawn, taking a sip of Coke from her glass. "Just nerves."

The topic of returning to Tabletop Mountain did not come up again that night or the next. But then, peculiarly, on Friday afternoon, Dawn called Steve at his office to tell him she'd changed her mind and wanted to go. Afterward, she spoke with her brother, Glenn, who agreed to babysit and it was decided: They would leave for Tabletop Mountain early the next morning.

Near Tabletop Mountain
10:30 a.m.

Moments before they departed Dawn put last night's scribbled note on their kitchen table. It read:

> 'Dear Everyone,
>   We feel we're being called back. God willing we will return. Our love for our children and family cannot be expressed in mere mortal words.
>
> > Steve and Dawn Hess
> > October 27, 1989'

Dawn was thinking back to those words she had written late last night, so somber and so true. It felt strange driving down Black Canyon Road in the Ford camper exactly as they had done one week before, she reflected. But there was more to their return than curiosity or the hope of vindication. Originally reluctant to venture back she, like her husband, had come to realize that until they found answers to the questions that plagued them, there would be no peace in their lives.

Steve cast a concerned look at Dawn as they passed Wild Horse Canyon Road and the Mid-Hills Campground. Silent for the better part of the trip, she was pensive and frightened.

"You okay?"

She nodded bravely.

"You're nervous, aren't you?"

"A little," she answered.

He eased his foot off the gas pedal slowing the truck as he got his bearings.

"Me, too," he admitted, "but also a little excited."

"I know. I could tell the other night at the dinner table when you first mentioned it."

Steve's eyes scanned the horizon. To the west lay the Providence Mountains; due east stood Woods Mountains and the valley they'd returned to see. They were near. Very near.

"Guess you know me pretty well," he remarked.

"Well enough to understand that you thrive on competition and probably see today as some kind of 'challenge'. Not me. They scare me, Steve, and I'm not ashamed to admit it."

He took the truck east onto Black Canyon Wash beginning their trek down the side of the mountain.

"What makes you think I'm not scared? I am. Terrified. But there's got to be more to this than just surviving. Hell, there are all kinds of crazies you read about who claim they've seen aliens. Just pick up the *National Enquirer*. But this really happened and if there's a way to prove it, I intend to."

"Well, I don't know," moaned Dawn. "When we first started out this morning I thought this was a good idea; to get to the bottom of this thing right here where it happened. Now I'm not so sure."

Steve held the wheel steady as the camper lurched forward over a crater in the road.

"I'm sorry, baby, is there something I can do?"

She guffawed. "Just tell me I'm not going crazy, I guess. You know that's what I feel like sometimes; that we've both gone stark raving mad together. That couldn't be what's happening, could it?"

He shook his head, solemnly.

"I don't think so."

"You know lately, I've come to believe they're in our house." She looked to him, head cocked to one side. "I mean, doesn't that sound crazy? But I swear when I was in the playroom with Bethany the other day, a wind came into the house and I knew they were there." She clenched her fists. "I just knew it."

Steve steered the truck to the right onto yet another unnamed trail as their trudge down into the valley continued.

"Why is that so crazy?" he asked. "If they can design and operate the kind of spacecraft we saw last week, who says they can't visit our home in La Mirada, California?" He grimaced. "No, I think they, whoever 'they' are, can do just about anything they set their mind to; and I think they've been doing it in the Mojave for a long time."

"Well anyway," she said, beginning freshly, "that's why I came. Not to prove anything to anybody else, but for me and our family. To prove that it really did happen and that we haven't both gone insane. Maybe then things will get back to the way they were. Maybe then we'll be able to live with ourselves again."

Their descent was almost complete and as the truck plowed through the foothills toward the wash beneath Woods Mountain, Steve could understand why indian tribes such as the Piutes and Mojaves considered this ground sacred. Truly, it appeared other-worldly; its stark, desolate terrain heightened by macabre volcanic formations that gave testimony to the violent turbulence that seethed beneath the surface.

Moments later the truck broke from the rocky, crater-ridden trail through to the dry desert lakebed that made up the valley.

"Here we go," Dawn muttered ominously, as if they were about to board a roller coaster.

Steve smiled weakly. Joking or not, something subtle but pervasive had changed the moment they entered, he realized. Like the victim of some terrible crime unable to face their attacker even in a courtroom, there was something of that feeling here for him and Dawn. He felt suddenly intimidated, even guilty. What was there that lurked so dark in his mind that caused him to feel this way? he wondered. What had they done to him?

"I'm going to pull the truck to the exact spot where we camped last week," he murmured.

Dawn simply nodded. The more traumatized of the two, her feelings of awe and horror at returning ran even deeper and more acute. To her, this was anything but a 'challenge' or contest of wills. Rather, she wanted to be rid of the experience; to exorcise whatever demons tainted them with these feelings of depression and remorse.

Still, there was one thing she knew immediately upon entering the valley. Her memories, blisteringly real, from the week before were anything but fantasy. What she and Steve had experienced was as real as the granite mountains that enveloped them.

Steve pulled the camper onto the belt of high ground near the base of Woods Mountain. He turned off the ignition, then sat for a drawn moment; a thousand impulses dancing like talons before him.

"Let's go," he said turning to Dawn, his expression solemn and determined.

The two walked some twenty feet from the camper to the firepit that Steve had dug the week before.

"Well, I guess we were here all right," he said in a

matter-of-fact voice. "Here's the firepit we built." He pawed at the burnt mesquite that lay outside the circle of rocks with the toe of his boot. "And here are the burnt branches that flew from it when I kicked the fire out."

"Yeah, I remember," she recalled in hushed tones. "When the little ones; the ones with the red eyes began landing all through the valley."

Steve and Dawn took several steps beyond the firepit. Slowly, their eyes raised up and directly across the desert some five hundred yards. There, towering like some stark and monolithic tribute to the stars and heavens above, stood Tabletop Mountain; its huge flattop configuration jutting some seven-thousand feet up from the desert floor below.

Dawn's body quaked involuntarily as she stared out across the desert basin. Here, spread before her was the landscape she had observed, trapped like a caged animal, hour after hour during their ordeal. And at that moment it all came back to her: The mothership with its white lights flashing, capping the entire nighttime sky above; the gremlins mocking their horror at every stage of the torture; the illuminated fugures, their enormous dark eyes ripping into their brain like daggers; the 'searcher' relentless as it canvassed every inch of the valley; the interminable sound of drilling resonating low and thunderous from deep inside the ground beneath the camper; the burning, sulphurous stench that permeated the air of the camper, so vile and smothering.

Dawn felt suddenly faint.

Steve grabbed her from behind as she staggered backward, mouthing soundless words that Dawn could not understand.

She tried to catch her breath, but couldn't. What was

happening, she worried as the visions she had experienced one week earlier came tumbling like boulders from beyond the recesses of her subconcious: Little Steven; the birth; the operating table; with blood gushing from his chest; surrounded by these 'things'...what were they doing to him? How helpless he looked! And now the rape. Her attacker; he's coming toward her; from a corridor filled with lights; she's lying on the ground; but it isn't the ground; and the rapist is not a man at all.

"Noooo!" Dawn screeched, her body arching in Steve's arms as she clawed and scratched, attempting to fight him off.

"It's okay. It's okay; it's me," he cajoled. "Dawn, it's me, Steve."

She stared into his face..

"Oh, I'm so sorry," she blurted, drawing him to her.

"No, no, it's okay, You had a flashback," he explained. "A series of them. I've been having them, too."

Dawn seized the intimation immediately.

"What were they about? What was in them?"

Still looking pale and faint, Dawn let Steve help her toward the camper.

"Are you all right? I mean, are you feeling better?"

"Yes. Yes. Now tell me," she insisted. "What did you see in your flashbacks?"

Steve opened the passenger side door of the truck, aiding her as she climbed into it. He was still contemplating a response as he reached for a Coke from the cooler, plucked off the tab, then handed it to her.

"I didn't really *see* anything," he began. "I mean, they weren't 'visions' like before; more like very powerful memories; recollections of things that happened last week."

"For instance?"

He hung his arm over the truck door.

"All of it. The spacecraft. The illuminated figures. But mostly the visions. Vivid recollections of that first sighting as a kid and the other...the one I told you about. No question those were the ones that came back the strongest."

"Were there additional events you remember?" she prodded. "Anything different from what we saw that day?"

He blew a long breath from his lungs; straining to remember.

"Lights...but not like the ones that were flashing," he uttered suddenly. "These are on runners. Like in that tunnel we talked about before." Steve cocked his head back, then shut his eyes very tightly. "Struggling..."

"What do you mean?"

"I don't know what I mean, Dawn. Just struggling; like in a wrestling match," he snapped abruptly. "Honest, that's all I can remember."

She nodded understandingly.

"How are you feeling?" he asked.

"Oh, I'm fine now." She dabbed at the beads of perspiration forming on her forehead. "It's the heat. And the memories." She shook her head, "And just the sight of this Godforsaken place..."

"We won't be staying. There is no evidence," he conceded. "Nothing physical, anyway. And that makes sense because I don't think they're physical beings. Not physical as we understand the word, anyway."

He stepped away from the door, then slammed it shut.

"Let's get the hell out of here."

He started walking toward the other side of the truck when Dawn called out to him.

"Steve?"

He turned.

"What is it?"

"We don't have to come back to find them."

"What are you talking about?"

"They're not here," she stated plainly. "I know when they're watching."

**Upland, California
February 14, 1990
9:35 p.m.**

It wasn't until several months later that Steve and Dawn decided to confide all that was happening to Dawn's parents. The experience was wearing them down.

For Steve, work had become unbearable. The sleepless nights and horrific nightmares continued, but as impairing was a feeling of guilt and isolation; as though he had done something terrible and held some dark and shameful secret that could be shared with no one. His powers of concentration at work had diminished and the man once described as "Mr. Dependable" at Southwest Construction now struggled with attendance and project follow through.

The weeks and months immediately following the experience had been even more devastating for Dawn. She became reclusive eschewing social contacts with friends and family for fear that they might discover what was happening. The dread she felt personally extended to the children, whose life outside the immediate family she vastly restricted, fearing that they might be injured or abducted by the creatures, whom both she and Steve believed watched them regularly.

With this sense of isolation and helplessness growing,

the two decided this was the time to share their experience. Dawn's folks, Ed and Bonnie, had been watching the children that day in Upland and they'd been invited for dinner, so it seemed a simple matter to stay on for coffee and the discussion they so desperately wanted.

Steve was sitting at the dinner table with Dawn's parents when she returned from the bedroom where Steve, Jr. and Bethany were sleeping.

Dawn joined Steve at the table. With coffee and dessert out of the way, Dawn sat forward in her chair, then began uncomfortably.

"Steve and I wanted to talk to you about something very private. Something we've told no one else about...no one."

Ed Larned, a strapping, soft-spoken man with short-cropped salt and pepper hair, sipped his coffee thoughtfully. Bonnie, his blue-eyed wife, quietly waited.

"Yes, dear. You don't have to be concerned about confiding in us," she encouraged. "No matter what it is."

"I know," said Dawn as ardently. "I've never had to be concerned about that, ever. Not even as a child. But Mom and Dad, something happened to Steve and me while we were in the Mojave back in October that's had a profound effect on our lives." She bit her lower lip as her chin quivered with emotion. "I know this will sound impossible to you, but we had an experience with extraterrestials."

Dawn's mom reached for her hand across the table. "You mean, you saw a UFO. Is that it?"

She shook her head, looking to her, then to her dad.

"There's a movie that came out not long ago called Communion. Do you know the one?"

"Yes, yes, I think so," Bonnie agreed glancing to her husband, who sat back taking a long sip from his coffee cup.

"Well, that's something like what's been happening to us; to Steve and me...I say happening because I'm not sure it's over yet."

Dawn's dad, a gentle, reserved man by nature, deposited his cup down into its saucer, then stared at both her and Steve quizzically.·

"Are you trying to say that you two were abducted by extraterrestials?" He cocked his head to one side. "Is this serious? Or are you two putting your mother and me on?"

There was a protracted silence during which he looked long and deep into his daughter's drawn countenence and mirthless green eyes. He had seen that look before.

"*They hurt you, didn't they?*" he asked finally.

She nodded woefully.

"It was horrible," she snarled, angry and hurt and crying now. "They came down from the sky, hundreds of them, and held us in the back of the camper."

"'They', the aliens?" asked Ed.

She nodded again totally absorbed now in relating the details of their encounter.

Patiently, and with true caring, they listened as Steve and Dawn told them what had happened and the depth of the emotions they were feeling.

By the time they had finished, it was nearly 1:00 a.m. and each was anxious to hear a reaction.

It was Ed, Dawn's dad, who spoke up, sensing the need to comment and somehow try to assess what they'd told him.

"I believe that you believe this has happened," he struggled. "Now, I'm no scientist, but what you're telling me defies everything I know. Space aliens, angels and gremlins," he shook his head helplessly. "I love you more than anything, Darlin', but there's got to be some other explanation."

"Do you think we're lying?" Steve asked bluntly.

"No," he answered unhesitatingly. "I don't. I think you and Dawn are telling us everything as you believe it happened."

"Isn't it possible you saw some experimental jets or rockets the Air Force might have been testing," Dawn's mom suggested.

"You don't understand, Bonnie," Steve objected. "We're not talking about two people looking up into the sky and thinking they saw a flying saucer." He waved his hand, palm up, from side to side. "No, no, it was...huge in scope. The spacecraft. The lifeforms, themselves, filled the entire valley!"

"Manuevers, then?" she offered.

Steve's eyes flashed.

"It wasn't experimental jets and it wasn't maneuvers," he said tersely.

"Okay, okay," Ed mitigated. "We know what you're saying, but please realize we're just trying to understand; to help you, if we can."

"Dad," said Dawn stretching across the table toward him. "You can help us by believing what we're telling you is the truth and that it really did happen the way we're telling you."

Her dad's sanguine complexion had paled over the past four hours. He was tired and perplexed.

"I hear you," he sighed, "but you know that part of it doesn't really matter. The important thing is that your mother and I see how upset you both are; we still love you; and we're going to support you in any way that we can."

"Your father is right, dear," chimed Bonnie. "We don't know much about these things. Who can say what's possible or not? But you'll always have us here for you, no matter how things turn out."

"Thank you," said Dawn embracing Bonnie and then her dad. "I love you both so much."

Steve stood. He shook Ed's hand, his disappointment palpable.

"One thing you said that I can't agree with. That it doesn't matter if it really happened or not. Well, it did happen," he said pointedly, " and it does matter."

Few words were exchanged during the short drive back from Upland to La Mirada. Generally, both were pleased at Ed and Bonnie's willingness to help and understand. But Steve remained vexed at what he knew deep inside was true: Dawn's parents could not bring themselves to actually believe the events as they'd described them were true.

*   *   *

Dawn and Steve lay in bed sleeping soundly that morning even as they sensed another presence in the room. That feeling of a stranger watching must have penetrated through to their subconcious because Dawn began tossing agitatedly while Steve awakened momentarily to glance at the clock radio on the nightable beside their bed.

2:38 a.m.

In dreams, Dawn imagined herself at a Sunday service with her mom in the familiar surroundings of their hometown church on San Antonio Avenue in Upland. Steve was locked in one of his proudest moments. A final minute interception during the East-West championship football game in 1986. All was as it should be for both, until the church singing stopped and the crowd's cheering came to an abrupt halt. Each of them stood, Dawn in church; Steve in his stadium...*watched*.

The churchgoers' eyes seemed cold and alien and even her mother turned to study this strange creature beside her. Steve was patted on the back and congratulated to the cheers of thousands at Bulldog Stadium, but now his teammates didn't know him and the crowd became scientists staring at him as if he was a curiosity and not a person at all...and in the dreams and around them always were the Eyes. Eyes like he'd never seen before; studying; probing inside both of their brains, until their hearts were pounding and their minds raced to the point of bursting.

Together they sat up in bed, their dream images suddenly merging into the reality of the dark, shadowy outlines that stood now at the edge of their bed. At that moment, it felt to them as if everything had stopped; time, even the steady throbbing of their elevated heartbeats. Steve again glanced at the clock radio; was he dreaming?

2:53 a.m.

A scream rang out from the children's room where Steve, Jr. and Bethany were sleeping! Dawn was the first to enter. She flicked on the light instinctively. Since the inception of their trip to the Mojave she had lived with the morbid fear that something horrible was about to happen to little Steven. Now, she stopped at the entrance to the room with her husband watching in awe and horror as their child stood sound asleep in the center of the room spinning like a top!

"You sons of bitches!" Dawn wailed running to him, then scooping him up in her arms. "You think you can do anything you want, but you can't! Don't you dare touch our children! Leave my child alone!"

Already Steve was beside her. "It's okay. It's all right now," he was saying as he guided them back toward the bed.

Little Steven, cradled in her arms, was only now beginning to awaken.

"Mommy, what's wrong?" he asked groggily.

"Nothing, dearest. Nothing. Everything is going to be fine," she soothed, laying him back in his bed, her heart still hammering. "Now you just go back to sleep. Everything is fine now."

Dawn backed quietly away from him as his heavy eyelids dropped. She joined Steve, who stood watching from the doorway, then flicked the light off. But the moment she did, the boy leapt up suddenly in his bed.

"Don't turn off the light!

"What? But why?" she asked, stunned.

"Because when you do, the little monsters come."

She sighed, "Steven, there are no monsters."

"But there are, Mommy! There are!" he insisted. "They're short and ugly; and they have red eyes..."

Steve and Dawn left the children's room that morning horrified to think that they were no longer safe even in the environs of their own home.

THREE

Barstow, California
May 4, 1990
1:30 p.m.

Steve sped his 1989 Honda Accord down I-15 headed for Barstow. The kids were with their uncle Glenn. This was the time he and Dawn had set aside to meet with Wolfy and Diane and Steve was apprehensive. How would his parents react? What would they think of him afterward, he fretted, if in the end they didn't believe their story and thought he had lost his mind?

The idea of telling Wolfy and Diane was risky, Steve reasoned, but what alternative did he have? Who, if not his own parents, would believe him after all? And these days, Steve knew, he needed someone like Dawn had in Ed and Bonnie with whom to share his feelings; to relieve the dizzying personal and business pressure that had been mounting daily.

Dawn sat beside him in the passenger seat catching up on some badly needed sleep. Since the incident with Little Steven, she felt a heightened sense of vulnerability that bordered on paranoia. Every creak in the house; every shadow in the night caused her to wonder if it was 'them' that had caused it. Oddly, though the insomnia and nightmares persisted for both of them, no tangible evidence of alien presence had revealed itself since February. But the terror remained. They lived with it and it affected every moment of every day.

The car phone sounded. Steve picked it up on the first ring.

"Yeah? Bob, how are you?" He grimaced. "Yeah, yeah,

I know I'm behind." He shrugged. "Hey! I'll get it to you, all right? No, absolutely not; but tomorrow," he promised. "Okay, okay, Bob. Yeah, goodbye."

Dawn's eyes opened slowly.

"Who was that?"

"Bob Jacobs, the stupid bastard."

Dawn frowned. "I wish you wouldn't talk like that. Here you are just baptised and converted to the Mormon faith."

He sighed. "You're right. I'm sorry. I'm just so far behind...on everything." He nibbled the cuticle of his index finger. "I'm afraid I'm gonna lose my job, Dawn."

She patted his knee.

"It'll be good for you to see your Dad."

"I hope so. Wolfy can be pretty cynical."

"You're his son. He'll believe you."

"Your parents didn't," he shot back.

"They wanted to." She pulled down the passenger visor, studying herself in the mirror. "And if he wants proof all he has to do is look at the circles around our eyes."

"Same dream?"

She nodded.

"They were around our bed again last night. Three of them. I could feel them. See them there like shadows in the night."

"I know," he mumbled somberly. "I see them, too."

He turned off at the Barstow exit.

"Are you going to tell them that part?" Dawn asked. "How we believe we're being monitored?"

Steve blew out a blast of air from his lungs.

"I just don't know," he confessed. "Early this morning I wondered whether we should tell them at all."

When they arrived at Steve's parents' house, Wolfy was

already on the porch waiting. He walked down the three short steps into the driveway.

"To what do we owe the honor?" he asked facetiously. "I don't think I've seen either of you in three months."

"It's been a while," Steve agreed, nodding.

The three of them entered the modest, ranch-style home. Diane stood in the center of the living room; a couch and family portrait directly behind her.

"Want a beer?" asked Wolfy.

Steve declined.

"How 'bout a nice glass of soda or wine for you?" asked Diane, stepping toward Dawn.

"No thanks."

Everyone looked nervous. It was Wolfy who spoke up.

"You know, I can probably save us all a lot of time." He gestured toward the sofa. "But first. Sit down." He watched as they did, then sat himself. "I think I know why you're here."

"I don't get it," retorted Steve.

Diane rushed to Wolfy's aid.

"What he means is..."

He gently placed his large hand over hers.

"I know what I mean," he argued. "People are concerned about you two. You look like hell. No one sees you anymore." He calmed himself. "Point is, that's why Dawn's folks and us talked. And better yet," he hesitated, "I think I may know the answer to your problem."

"What problem?" Dawn asked.

"The UFO's; the extraterrestials; all of that," he stated directly.

"We're listening," said Steve.

"Steve, you know there are quite a few military installations spread right around the area of the Mojave you were in," he began. "Twenty-nine Palms. Fort Irwin. Nellis and the Naval Ordinance Test Station. My first guess is that you two saw some experimental craft in operation."

"No," said Dawn. "Definitely not!"

Wolfy held up his hand. "Hold on. Let me finish. There's another possibility and that is you witnessed military maneuvers. Hellicopters, desert hover craft, soldiers with night vision goggles that glow red in the night."

He'd gotten their attention.

"There's a Desert Warfare Training Center in Fort Irwin. In a full-blown exercise they might even use gas canisters. Remember, there's a lot of activity going on in the Gulf right now."

Wolfy had made his point.

"No, no," Steve objected. "The type of aircraft we saw doesn't exist in any military, Wolfy. There was one craft," he tried to estimate, "one hundred yards across."

"And the Beings, themselves," Dawn interjected. "Wolfy, they just weren't human."

"Anything else?" Wolfy inquired.

"Yes," said Dawn. "It still goes on." She became suddenly agitated. "They're everywhere, watching us, monitoring every thought, every emotion. And we see them. Never really close up and clear like in the desert that night, but we see them." She squinted as she refined her thought. "Almost where we don't see them. Movement without anything being there. Shadows at the edge of our bed at night. And always these big, dark insect eyes!" Dawn stopped abruptly as if she had caught herself drifting beyond the outer fringes of sanity.

"We know what you mean, dear," said Diane, nodding politely. "We understand."

Wolfy sat back in his chair shocked at the conviction and level of intensity in Dawn's voice. Then, he spoke very softly.

"There's a last possibility that I should mention," he said reluctantly. "It's not so unlike the last explanation and I wouldn't mention it except that I once heard of a man who claimed he'd been part of a top secret 'psy-ops' program in the mid-80's. A good, decent fellow. College graduate, Georgetown, I think; with a wife and two kids." He sat forward staring both Steve and Dawn in the eye. "When they got through with him, he had to be toilet trained 'cause his mind had been reduced to a complete blank." Wolfy paused a moment to let his idea settle. "They got a bunch like that at Fort Irwin Desert Training Center." He dropped back in his chair. "Now, I'd like you two to just think about that."

Not a word was spoken for at least thirty seconds.

"So you're saying we may have run into one of these 'psy-ops' divisions," surmised Steve, "and that we were drugged, or gassed or whatever, then held captive by our own military while the maneuvers continued."

Wolfy placed the palms of his two hands over his belly.

"That's what I'm saying. And if you're asking me to choose between monsters from outer space and drugs and war games; the war games win every time."

"My sweet Jesus!" gasped Diane. "You don't think..."

"Why not?" asked Wolfy. "They've done it before, the bastards."

Dawn turned to Steve. "That could explain the flashbacks..."

"You've got to be kidding!" protested Steve. "Do you think I can't tell the difference between what we saw and guys wearing night vision goggles?"

He looked to Wolfy.

"And let me ask you this. If we were drugged and hallucinating, how is it possible for two people to hallucinate the exact same things, at the exact same time."

Wolfy gazed at his son, deadly serious.

"I don't know. But what I do know is that you two can't go on like this." He motioned to the outside through the living room plateglass. "Look around you. It's spring out there. What happened to you was more than seven months ago! Try to relax. Enjoy the flowers and trees and your children. If things don't improve by summer, I'd suggest you both get some professional help."

It was Dawn who spoke up now in a steady, cadenced voice that belied her emotions. "I'm sure you're both worried about us," she cajoled, "but Steve's right. We saw them. We smelled them. We felt them around us. At exactly the same moment in exactly the same way. No," she reasoned, shaking her head from side to side, "this was no military maneuver and it was no hallucination. In my heart, I know what we witnessed was not of this world."

Steve and Dawn had a late lunch at his folks' house, then drove back home under the late afternoon sun. They picked up Steve, Jr. and Bethany from Glenn's house, actually feeling better since their visit. Though they didn't believe Wolfy's theory, it was the most viable alternative they had come across.

Dawn joked that even being kidnapped, drugged and brainwashed by our own military seemed less sinister than what they believed they'd actually been up against!

La Mirada, California
June 12, 1990
3:05 a.m.

It was four weeks later in the dead of early morning that the dreams recurred with an intensity unlike any of the others. In his, Steve saw himself once again back in the camper, trapped and screaming, while the illuminated creatures encircled him. He was screaming, but there was no sound...And then he saw a flashing image. It was of a long narrow tunnel with lights running along the sides...but then it was gone again and then he saw a Being who was all white, the color of the tunnel lights...Again he saw a flashing image and it was of several Beings like that...they were trying to restrain him...Then the vision was gone and only the first figure remained...

...THE BEING IS FOUR FEET TALL...HE WEARS A WHITE LUMINOUS UNIFORM WITH AN UPTURNED ARROW ON THE CHEST...HE HAS NO FACIAL FEATURES...NO MOUTH OR LIPS...ONLY SLITS...HE PASSES DIRECTLY THROUGH THE WALL...HE STANDS BEHIND THE HEADBOARD...HE STAYS THERE...PASSES HIS THREE LONG FINGERS OVER DAWN'S FACE...OVER AND OVER...STEVE CAN ONLY SEE THE HAND...HE OPENS HIS EYES AND DIRECTLY ABOVE HIM IS THE FACE OF THE WHITE BEING...'YOU'RE DREAMING' HE HEARS THE VOICE SAY...THEN HE LOOKS TO DAWN'S FACE AND SEES THE BURN MARKS ITS FINGERS HAVE LEFT!!!

"Hey, wait a minute!" Steve said aloud. "This is no dream!"

He turned to Dawn and gasped. He was awake. He knew he was awake and Dawn had the burn marks all over her face!

Dawn's eyes opened. The expression on Steve's face scared her.

"What? What is it?"

"Go to the mirror," he urged her. "Go to the mirror now!"

Dawn walked to the wall mirror above the chiffonier.

"My God, what is this!" she screamed. "Steve! Steve!! What have they done? They've burned my face!"

"That's all right," he said holding her close to him. "It's gonna be okay..." He eased away from her. "Now you just stay right here, sweetheart, while I go get the camera."

Steve came back breathless with a Polaroid in hand. Dawn was horrified; hysterical as he snapped photos one after the other; still trying to calm her and himself as he clicked away.

"Just take it easy. Everything is going to be okay." *Flash!* He sucked for air.

"That was him." *Flash!*

"In the tunnel with lights." *Flash!*

"He and three others are who I was struggling with." *Flash!*

"That night in the desert." *Flash!*

Steve took five photos of the burn marks on Dawn's face that morning. When developed, they showed the marks resembled patterns and symbols; a thunderbolt on her right and on the left the imprint of three distended fingers wrapped around Dawn's face from behind.

* * *

Later that morning Steve awoke to the sound of Dawn sobbing hysterically as she stood before their full-length bedroom mirror observing the burn marks on her face.

Neither had any recollection of what had occurred hours earlier and so he, like Dawn, was shocked to discover them.

"Look at what they've done to me," she wailed. *"Look at what they've done to my face!"*

Steve walked to her. He touched her right and then her left cheek studying the peculiar markings so clear, yet so inexplicable.

"And here," she raved, waving the five photos in his face. "Look at these!"

Steve flipped through the photos.

"Polaroids." He shook his head incomprehensibly. "Did I take them?"

"You must have," Dawn hissed. "You must have taken them this morning. But look! The photos are blackened. And they made us forget again. Don't you see what they're doing?"

"Baby, I'm so sorry," he whispered, holding her. "Does it hurt? Should I call a doctor?"

"No!" Dawn bellowed. "I don't want anyone to see me like this. Do you understand that?"

Somberly, he nodded, watching as his young wife felt the side of her face with her fingertips.

"They don't look like they'll scar or anything, if that helps," he soothed, trying to understand what might be going through her mind.

"It's not that," she seethed. "It's not just that." She swung around to him. "This is our house. Do you understand that? Our house!" she repeated. "With our children. And they think they can just do any goddamn thing they feel like!"

Dawn fell into his arms like a marionette whose wires had suddenly been severed, sobbing uncontrollably.

"I'll stay home from work today," he comforted. "I'll take care of the children. You just relax. Call your mom. Talk to her. Take as long as you want 'cause Dawn, I swear to you, I'm going to get to the bottom of this. I'm going to get there," he vowed, "if it kills me."

When Bonnie arrived at their home, she was shocked to see Dawn's shattered psychological condition and the mysterious markings that covered her face.

The welts remained visible for another twenty-four hours when suddenly they faded, then disappeared entirely.

Thousand Oaks, California
August 23, 1990
3:30 p.m.

Paul Moran had been a party animal in college. Standing 6' 1" and weighing 190 pounds, he was a small defensive lineman at Redlands, but what he lacked in size he made up for in heart. Aggressive on the field and wild in the nightclubs, he was also a guy who wrote poetry offseason. Since college he had settled down, going into business for himself, but always through whatever changes had occurred in his life, he and Steve Hess remained best friends.

This afternoon the two sat in a darkened corner of the British Pub, Moran bunched over on the table mesmerized by what Steve had told him.

"I don't want to burden you," pled Steve, "But who else do I go to? My own parents don't believe what I'm saying..."

"Yeah," Moran agreed, locked in thought as they stared at one another, eye to eye, for a drawn moment, "but I believe you."

Steve's eyes narrowed.

"You mean that?"

Paul smiled, nodding.

"Hey, you're the 'Duke'!" He slapped Steve's shoulder from across the table. We were roommates all through college. If I can't believe the things you tell me, I can't believe anything."

Steve's eyes were still wary.

"So, you don't think I'm crazy?"

"Hell, no!" he pledged.

Steve withered.

"Well, Dawn is starting to think we've both lost it and to tell you the truth," he confessed, "I was starting to agree with her."

Moran was shocked.

"No, no," he argued, dismissing the notion out of hand. "I don't know a lot of things, but you don't have to be Albert Einstein to realize that 99% of what we take for granted as real is either a bunch of scientists kidding themselves or delusion."

Steve smiled for the first time in a week, then took a sip of Guinness.

"Five hundred years ago people thought the world was flat. That was 'real'," Moran shook his head, "but it wasn't. Three hundred years ago Galileo was called a heretic because he didn't believe the planets revolved around the earth. That was 'real', too," he shook his head, "but it wasn't." He looked at Steve squarely from across the table, his dark eyes and chiselled countenance intent. "I say, the next one to fall is Newton!"

Steve shook his head, laughing. "Moran..."

"No, no, I mean it. I read a book called *Chaos* by James Gleick that proves Newton was dead-ass wrong in his theories about gravity." He took a pull from his stout. "If that goes, my friend, the book is rewritten on *everything* that's possible and impossible!"

Moran sensed the inner sadness that his buddy harbored and how deep-rooted it must be.

"How is Dawn?"

Steve's eyes lifted from the tabletop. He took a gulp of

Guinness.

"Not good."

"This?"

He nodded. "Paul, you don't understand how deep a thing like this runs. We don't sleep at night for fear of what they'll do to us; we worry about our kids constantly; Dawn hasn't had her period in months and though she doesn't say it, I know she believes they've somehow altered her." He fluttered his hand at the ineffable. "I don't know, taken her ovaries, I think."

"When would they have done that?"

Steve shrugged.

"We don't know," he explained. "The part I told you about in the camper lasted eight hours. We were conscious and pretty much aware of everything for the whole time." His brow furrowed. "But when we slept, Dawn thinks we were already on board the craft in some kind of holding area. That may be when it happened. We have no way of knowing."

"And you've told no one else about this?"

"Just our parents; my brother, Ron, and Dawn's, brother, Glenn."

"Because you're afraid no one else would believe you?"

"That's right. I like my job, Paul. I'm already in enough hot water without this kind of thing. Everyone would think I was crazy," he snickered, "or on drugs."

"So you and Dawn just continue to carry around this incredible dark secret?"

"We live with it," Steve retorted.

Moran toyed with his beer glass.

"My God, Steve, is there anything I can do to help?"

Steve stared at him deadly serious as he brought his glass to his lips.

"That's the point, old friend. There ain't nothin' nobody can do."

It was an hour and a half drive back to La Mirada from Thousand Oaks on the 101, so Steve had some time to try to put his and Dawn's situation in perspective. Certainly, Paul Moran's words of encouragement had helped because, God knew, he felt alone these days—so did Dawn and it was eating away at her.

In some ways, that was why he'd converted to her church, Steve ruminated, to be with her in something she trusted, but as important, to give him something to fight back with. He needed some weapon against them; to stave them off and probably it was religion because in his heart of hearts he still could not comprehend whether the creatures were consumately good or consumately evil.

By the time Steve got home, it was after six and he'd missed dinner. Dawn was annoyed at him for that, but more for missing work. So, with an air of tension in the house, he munched a cold lamb chop with a glass of milk, helped Dawn give the kids a bath, watched the last half hour of a TV movie, then retired for the night.

La Mirada California
September 10, 1990
3:35 a.m.

The episode in June involving the burn marks on Dawn's face had jarred the Hesses to the very core of their existence. Now it was clear that nothing was sacred; nothing was private; and the scope of alien involvement in their lives was total. More, it was obvious that there was much that they did not know. About the enigmatic markings on Dawn's face, about the photographs they surmised had been taken by Steve; and about their original experience in the Mojave itself.

Still, bits and pieces of information concerning that period began seeping through from their subconcious like random pieces of a puzzle; mostly in images and always in half-remembered dreams such as the ones both experienced on the morning of September 10, 1990.

The images were clear, but disconnected for Steve, who pictured himself in the Mojave engaged in a deadly struggle with four of the white Beings who pulled him and Dawn from out of the camper. Drugged, or somehow altered mentally, he lashes out at them, flailing his massive arms in an effort to resist.

Dawn is beside him being led by two others. She reaches out grasping for his hand as she is being pulled away, but it's no use for now he, too, is under their telepathic control, ambling stiffly away from the camper through a long tunnel of lights.

He and Dawn are separated now. And in the dream Steve is again thrown back to the struggle...the struggle to resist...with Dawn's arm extended, her fingers stretching out for his help, but grasping air!

For Dawn, the images only began at that moment as she envisioned herself already alone being led by a white, shimmering Being down a tunnel; walking, but not walking, more like floating. Is she dreaming or drugged, or what is she? Dawn finds herself wondering. And where is Steve? And why is she so frightened if everything is okay like they keep telling her? But it isn't okay, she knows, because Steve is gone and she is floating down the tunnel with blue lights, until they arrive at a white room.

She enters. The table. The operating table is silver and shiny like chrome. With the round light that stares down at her like a huge, round eye.

Dawn dreams about the light and about waking from the dream to see the large, flat, insect-like face of the alien examiner as it stares deep into her panicked eyes!

Dawn's head jerked from off of her pillow in time to see the fleeing shadows around them disperse.

"*Turn on the lights!*" she cried out to Steve, who was already awake, chest heaving, beside her. "*They're here, aren't they?*"

Steve sat up. His vigilant eyes scanned every corner of the room.

"*Yeah,*" he answered breathlessly. "*They're here in the room now!*"

He got out of bed, then flicked on the lights. As was generally the case, the telltale signs of their presence lingered, if only for a moment: the coldness; the subtle, but unmistakable odor—musty and stale—like a cigarette had just been extinguished in the air around them.

But the shadows were gone now. Submerged; rarified; who knew what? Only that the terror and sense of violation remained.

Steve walked back to the bed, then sat on the edge beside Dawn.

"Do you remember any of it?"

She shook her head in frustration.

"Eyes. Their eyes, like before. A table, I think. But that's all. You?"

"A sense of confrontation; physical confrontation with them that morning in the desert." He attempted to rein in the memory, or the sensation the memory had created, but came away with something else. "It's almost like they're trying to tell us something. Do you sense that? Not malevolent like I thought at first. Straining. Like us. To communicate."

Dawn dug the heels of her hands into the sockets of her burning eyes; massaging them; perhaps attempting to open them to a second kind of vision.

"It's been too long, Steve," she admitted, as if in defeat. "Maybe Wolfy's right about getting professional help."

He nodded wearily.

"Maybe, but I just keep thinking something's going to come of all this...something, though I swear I can't say what it is."

Meeting with Paul Moran
Westlake Village, California
November 17, 1990

It had been over a year since I'd seen Paul Moran. For a five year period or so we'd worked together at a major New York Corporation. Now he was in his own business situated on the West Coast where he'd been raised and educated, so that our dinner on the patio of the Orleans-West was something of a reunion.

Certainly, it wasn't a difficult situation to enjoy. The good company of an old friend, a magnificent California evening and a casual dinner overlooking the lake with its sailboats and lush shoreside homes.

Tall and trim, Paul's dark hair, boots and jeans made him look like a cowboy as he stretched out on the near empty veranda sipping from a glass of California Chardonnay.

A classic easterner by comparison, I drank from a bottle of Heiniken and was dressed in tailored slacks, a collared shirt and Bostonian loafers.

Perhaps it was the contrasts as much as the similarities

between us that always made the conversation flow easily and tonight was no exception.

A bit of a poet himself, Paul had always been interested in my writing and asked about it.

"Working on any books these days?"

I shrugged.

"Preliminary."

"Non-fiction?"

"Yeah, as a matter of fact. I'm thinking about two projects," I explained, "but haven't decided between them."

Moran smiled broadly.

"Tell me about them."

He knew he wouldn't have to ask twice.

"Okay," I began. "One would be a picture book. Location shots of sites around the world where miraculous occurences were said to have happened: Lourdes, Fatima and lesser known, more current ones. Of course, there'd be text that goes along with each. I was going to call it *Miracles*."

"And the other?"

"This one is about a former POW I met recently. His time imprisoned and tortured, but more specifically his rehabilitation afterward. The man went through hell in southeast Asia, but oddly the after affects of trauma like that can be more devastating than the actual situation."

The coupling of the two topics had set Paul's mind working.

"That's the one I like," he said taking a sip of wine. "But how would you go about researching something like that?"

"Well, extensive interviews with the man himself, of course. But a former roommate of mine at Georgetown is a psychiatrist these days. I'm hoping he'll put me in touch with another alumnus who's Director of the National Trauma Center in Washington, D.C. I'd go to him for background."

"And that's all this 'Center' treats; victims of trauma?"

"Yeah, that's what Bernie Vittone specializes in," I elaborated. "The after affects of trauma. Victims of serious accidents, violent rape, former POW's. He's supposed to be one of the best."

"If someone came to see him, let's say, from out of state, suffering from true terror after an experience they'd had, do you think he could help?" Paul asked.

"That's his expertise," I was quick to answer. "If he couldn't help, I'm sure he'd refer them to somebody who could."

Paul was pensive, I could tell, very carefully weighing two allegiances, one against the other.

"I know a true story that's better than either of the ones you're researching," he confided as grave as I'd ever seen him. "It was told to me in strict confidence by my closest friend, but because I trust you and think you may be able to help, I'm going to tell you."

"I'm listening."

"You've read about these alien abduction cases?" he asked.

"Sure. Bud Hopkins and all that. Go on."

"What if I told you I knew two people who were held captive and psychologically brutalized for periods of time, not unlike POW's?"

"I'd say you were putting me on."

"What if I told you that to my knowledge there's never been a story like this." He was emphatic. "Ron, these people witnessed alien activities on a massive scale while being held captive for more than eight conscious hours!"

"Where did it happen?" I asked.

"The Mojave desert, late October, 1989. But, as you say, for them the aftermath has been more hellish than the experience itself."

"Sounds like a nightmare," I blurted, affected by the story and Paul's obvious concern.

Paul folded his hands before him on the table, then sat forward in his chair.

"Ron, if I supplied you with the background on the case and let you meet these people to hear what happened first hand, would you try to arrange for Dr. Vittone to spend some time with them to see if he could help?"

"You mean do a book on them?"

"If it was of interest to you and they were willing."

I stared out across the lake. A Sunfish sailed by in the distance. My eyes drifted along with it as I considered the idea.

"The story doesn't sound like the kind of thing I do, Paul. But, of course, I'd be willing to meet them. With regard to Dr. Vittone," I shrugged, "this is a friend of a friend, so I make no promises."

Moran's eyes locked with mine.

"If you meet the Hesses, Ron, and hear their story, you'll want to write the book. From my point of view, frankly, I see a fit for you and for them. You'll want to write their story and they may just be ready to tell it."

This initial commitment to meet with the Hesses made out of politeness, evolved considerably as we discussed their experience in more detail.

By night's end, I was intrigued. My plans were to contact Dr. Vittone at the National Center for Trauma in Washington, D.C., the next morning.

La Mirada, California
November 20, 1990
8:30 p.m.

Steve Hess sat in the study of his home rapt in conversation with his longtime friend.

"I can't believe you told someone about this, Paul," he said angrily. "You promised; gave me your word it was in confidence!"

"I know," Moran acknowledged. "In retrospect, it was wrong to do and for that I apologize."

Steve Hesses' temper quelled. It was difficult to stay angry at Paul; primarily because he usually had your best interest at heart and Steve recognized this.

"But I still think you should meet him."

Steve's response was immediate.

"Do you realize that since this began more than a year ago Dawn and I have told no more than a handful of family members and you?" He rose from his chair. "Why on earth would we confide our innermost experiences; our most intimate feelings to a stranger?"

"For one thing, he's a stranger to you, but not to me and I'm telling you, Steve—it would be healthy."

"*Healthy?*" Steve spat the word back at him.

"Yes, healthy on two counts," Moran explained. "First, this account is true and the rest of the world needs to know about it. You've said that yourself, that you thought there was some reason you and Dawn were chosen for the experience. Maybe this is the time to step forward. And there's another."

"Yeah?"

"Ron Felber feels he can arrange some time for you with the Director of the Center for Trauma in Washington,

D.C. if the meeting between the three of you comes off. Maybe there'd be a book and maybe there wouldn't, but Dr. Vittone should be able to steer you and Dawn in the right direction at least and that would definitely be worth something."

"Paul, I appreciate your interest," said Steve emphatically, "but I just don't see it. What's this doctor going to do for us?"

"How about uncover the truth?"

"He's in D.C., Paul. He can't treat us from there. Besides, what's happening is happening. It's real."

Moran leaned forward in his chair.

"Then, prove it."

"How?" Steve snapped back. "How can anyone prove what no one will listen to!"

"Look, when we talked last time you told me that you and Dawn worried you might be crazy. Did you mean that?"

Steve looked him square in the eye. Slowly, he nodded.

"The first thing he'll do is a psychological evaluation. At the end, he'll know and you'll know if either of you suffers from psychiatric or organic illness." Moran was issuing a challenge. "He'll help you uncover the truth."

"What if we don't?"

"If you don't, I suppose people have got to start looking real close at what's been happening around this place."

"And if we do?"

Moran took a deep breath.

"Then you know you need help and we try to get you some. Steve," he argued, "I'm your best friend and I'm telling you, you just can't go on living like this."

Steve deliberated.

"Dawn will never go for it."

"Why?"

"For one thing, she's reluctant to tell anybody. For another, she's become very involved with the church. Frankly, I'm not sure she looks to medical science for the answer."

"Has it helped?"

"Internally. It's a source of strength. But no other way."

"Then, what do you or Dawn have to lose? This might not be the answer you're looking for, but it's the first step along the way."

"What would we have to do? What would Dr. Vittone want from us?"

"If you meet with Ron he'll tape interviews with you and Dawn. He'll ask questions about you as individuals as well as details about your Mojave encounter and what's followed. Copies of those tapes will go to Dr. Vittone in Washington. If he thinks he can help, he's agreed to see you."

Steve considered what Paul had said about wanting to 'uncover the truth'. He found the phrase appealing because in reality the truth was hidden from them.

"I'll talk to Dawn and see what she thinks. But yeah, I think I'd like to go for it!"

Westlake Village, California
November 23, 1990
9:30 p.m.

Paul Moran was watching John Carpenter's "The Thing" on video with his wife, Laura, when the phone rang. She answered, still munching a handful of popcorn.

"Paul, it's for you..."

He took the receiver.

"Hello?"

"Paul, this is Steve. Let's get the date set up with Ron Felber. We've both decided to step forward."

"*All-right!!*" the ex-football hero exclaimed.

Interview with Hesses
Norwalk, California
December 13, 1990

My next visit to the West Coast was about one month later. Prior to my arrival, Paul Moran had arranged several telephone conversations between the Hesses and me; all of them successful in that an instant rapport seemed to exist, but even to our first meeting, there was a hesitancy on their part to share what was obviously a very traumatic and personal part of their lives.

On the evening prior to our interviews, Paul introduced us over dinner at El Torito, a local restaurant, so that by the time they arrived at my hotel room at the Marriott Courtyard, the ice had already been broken.

By anyone's standards, the Hesses were a handsome couple. Well groomed and neatly dressed, Dawn was tall and trim with shoulder length chestnut hair and a face that exuded both warmth and intelligence. Steve was a strapping man standing 6' 1" and weighing 225 pounds with dark brown hair and thoughtful, deep-set eyes.

Both were nervous at the onset, but opened up once the conversation shifted to familiar topics such as family, athletics and the ordeal they had been through. They began by explaining how they came to be in the valley, then went on to describe the first sighting over Woods Mountain leading to the initial display of the nine shining objects in the desert sky above them.

From the onset of their account it was apparent to me that the encounter had deeply affected them. The attention to detail, the interaction between them, the feelings they attached to events as they transpired—all pointed to the fact that these were two well educated, intelligent and well-balanced individuals who had been stunned by the situation that enveloped them that night and the morning to follow.

While describing the invasion of the valley and their torturous hours in the back of the camper surrounded by the illuminated figures, their word choice, cadence and inflection spoke volumes beyond what they were saying. It was obvious that both had lived with a sense of terror that few individuals will experience in their lifetimes.

As significant, perhaps, were their feelings about the aftermath of their encounter and what it meant to them today.

"So tell me, has the experience affected you in any way since?"

"Every single day," Dawn answered directly. "We're scared and, I guess, confused. On the one hand, we'd like to think it was positive, but. well, maybe it's both."

"How is it positive?"

"I think we came away with another sense. As if we used a part of our brain that we've never tapped before and sometimes in the dark it's like we're seeing another reality; something behind this reality," she explained. "People or spirits behind what is our world. But we don't know how to get back to using it again."

"What would you use it for?" I asked.

"To communicate with them directly."

"You say that as if they communicate with you..."

"They do," she stated bluntly. "And maybe that's the

dark side of our experience."

"Please explain."

Dawn looked to Steve.

"What happened to us isn't over," he elaborated. "In fact, right now I can feel the chills through my body knowing that they're listening; monitoring what we say and do." Steve, who had been the calmer of the two throughout the past six hours, seemed suddenly angry and sullen. "At night you wake up afraid and feeling like somebody's watching you; standing at the edge of our bed. Black shadows. Movement without anything being there. Monitored?" He laughed sardonically. "Definitely."

"So, if these Beings are here watching, monitoring what you're dong, why would they do that?"

"That's the confusing part," retorted Dawn. "In some ways I think we're meant to do something and we're still waiting to find out what it is. Every day we wait with anticipation wondering when and where it's going to happen and what to do about it..."

"We try not to let it preoccupy us," Steve interjected. "I mean, I've got to work and not let this affect me. But it's like something needs to be said or something needs to be done."

He looked up to me. Tears were in his eyes.

"I understand," I said in a quiet voice. "Are you afraid?"

Steve's jaw braced.

"If they want to take me and kill me, they can," he stated plainly. Then, softly, as if he could not comprehend the notion, "but the children."

"How do you mean?"

"Our house creaks," Dawn said in a complete non-sequitor. "The water pipes," she explained, "the building settling; but it's more than that. Little Steven sees them. He talks about the little monsters and once not long ago when I got him one of those day-glow sticks as a present, I thought he was going to pass out he got so upset. It's the illuminated ones. That's all he could think of when he saw it!"

I settled in my chair, uncomfortably.

"So, if it's not necessarily good or evil...what is it that you're experiencing?"

There was deep and heavy silence.

"We don't know," Steve said at last. "We're just two people. Not anyone special. I can only tell you what's happened to us. I can only tell you that it means something and that I don't think they're through with us yet."

The interviews lasted two full days. By their conclusion, I was personally convinced that Steve and Dawn Hess had lived through and were living a fantastic and important experience that needed to be documented.

The interviews, recorded on six two hour tapes, were forwarded three days later to Dr. Bernard Vittone at the National Center for Trauma in Washington, D.C.

La Mirada, California
December 18, 1990
3:35 a.m.

Steve was awake; eyes wide open, staring at the ceiling as he lay in bed beside Dawn. All around the bed, the shadows stood, watching. He knew it; and it bothered him no longer.

He reached across the bed, brushing the side of Dawn's cheek with his forefinger.

She, too, lay on her back, staring straight up. "They're here, aren't they?"

"Yes," he answered.

"I can feel them around us. I can see them, too."

Neither moved a muscle.

"The kids?" she wondered.

"Asleep."

"But are they all right? Are 'they' in the room with them?"

"No. Not tonight."

He rolled over toward her.

She stared deep into his eyes, plaintively.

"This isn't right, you know."

"You know what they want."

She nodded, peeling her nightie away, laying there for him to take.

They made love.

He rolled back to his side of the bed again, the Beings still encircling them.

Dawn's eyes remained open.

"I don't want to be pregnant," she said, an edge of fear rising up in her voice for the first time. "Do you want me to be pregnant?"

"No," he answered, "but you are."

Chatham, New Jersey
March 10, 1991

After an exchange of tapes and letters and numerous telephone discussions, Dr. Vittone agreed to clear one full day at the National Center at which time he would undertake a complete psychological evaluation of the Hesses.

I contacted Steve and Dawn. The date was set for Monday, June 10, 1991.

FOUR

National Center
Washington, D.C.
June 10, 1991
2:45 p.m.

The Hesses sat uneasily in Dr. Vittone's office at the National Center for Trauma in Georgetown. It had been a grueling day, especially for Dawn who was now pregnant. Over the last five hours the doctor, already familiar with the account, had them retell it from start to finish, often asking three and four questions about some detail or feeling they had described. In this way, he was able to explore the depth of their experience and gain insight into their logic and emotional make-up. Dr. Vittone's objective was clear: to determine if the Hesses suffered from any psychiatric or organic illness that could account for their experience.

The doctor entered. Fortyish, he was lean and wiry, a little taller than average with a brown mustache and a reserved, personable demeanor.

He sat down in a comfortable leather chair directly across from them.

"One thing I can say is that you aren't crazy or psychotic in any way," he began with a smile. "I also don't see any indications your experience involved psychotic hallucinations."

"Could you explain that?" asked Dawn.

"Certainly," he answered. "There are two types of hallucinations, visual and auditory. Visual tend to go with psychotic reactions like schizophrenia or manic depression. Auditory occur in response to some phsyical stimulus such as hallucinagins, alcohol or brain tumors." His deep-set, intelligent eyes attempted to engage them. "Point is, they never occur together. For two people to have both auditory and visual hallucinations, and then to have them be identical?" He shook his head. "Impossible."

"So you're ruling out the possibility of mental illness?" Steve ventured.

"That's correct," he said, nodding thoughtfully. "Neither of you demonstrated any symptomology consistent with psychiatric or organic illness before or during the incident you described here today."

"What about drugs?"

"Same holds."

"How about extremely sophisticated drugs or chemicals that the military might be testing for psychological warfare?" he persisted.

"No. That doesn't change it," the doctor answered. "For you to get visual hallucinations means you'd have to be going through the same physical phenomena. So, fine, let's say that both you and Dawn have been exposed to or somehow ingested the same hallucinagin at the same time. It's true you'd both be hallucinating. Then, let's say you look up at the stars. It's possible that you'd both have visions of apparitions. But for them to be identical and doing the same thing? That could never happen."

Dawn appealed to him.

"Doctor, if it's not mental illness and it's not drugs— what is happening to us?"

"That I can't answer. But I do have some observations about your experience and the symptoms you describe that may be helpful."

The doctor slipped a notepad from out of his jacket pocket.

"Throughout the account you make reference to a fog or mist that you believe the creatures used to settle you down and control your breathing when events got out of hand. Is that correct?"

"Yes," they agreed.

"I have an alternate interpretation that views the 'fog' as a panic reaction system." He gestured with his pencil. "You both talked about pressure in the chest, hyperventiliation, cold flashes and the mist which we'll call haziness of vision. All of these are associated with panic caused generally by overwhelming, traumatic events."

Dr. Vittone turned the page of his notepad.

"Similarly, when you describe the creatures "playing with your mind" and subjecting you to "psychological torment", there is also an explanation." His eyes raised from the notes. "When people are put in circumstances where traumatizing, terrifying events are going on around them, they put up defenses," he explained. "One of these is by disengaging from the terror of the situation by basically pretending you're not there. These "visions" may have been your way of doing that. Now, the fact that they were not pleasant in all cases suggests to me exactly how frightened you must have been."

Again, he turned the page.

"The final point I'd like to make involves the reactions you've described most recently, after the experience. Fear, depression, feelings of violation and helplessness, the sense that you are being watched or monitored. Once again, these are the symptoms people experience after severe trauma." He gauged their reaction, then went on. "In other words, I can't say if you're being watched as you claim, but I can say that many people would feel that way after going through an experience such as you've described."

"So you believe us?" Steve ventured.

"I believe that what you're experiencing is a normal reaction to highly unusual and traumatizing events outside the realm of normal, human experiences."

"We appreciate that," offered Dawn. "But what can we do now?"

His eyebrows arched.

"Good question. If you were local, I'd suggest you spend some time here, but there's more to your situation than the treatment of trauma." He tapped his pencil against the palm of his hand. "In both accounts, there's clarity of detail and, at least, comprehension of what's happening, until the very end. Dawn talks about falling asleep "fully clothed" and you, Steve, about falling asleep "instantaneously" with the creatures still there. This is the part that troubles me most," he concluded, "The four hours between your going to sleep so suddenly and waking up the next morning."

Dr. Vittone stood, then walked to his desk. He opened the top drawer and produced an address book. He turned the pages searching for a name, then jotted it on a piece of paper.

"You seem to be people who need to know the truth about what happened that night in the desert. If that's the case and you want to pursue this, here's the name of a colleague who specializes in retrogressive hypnotic technique for victims of trauma. But I must warn you, the procedure can be emotionally jarring. Events plunged deep into the subconcious are generally there for a reason."

The doctor handed the slip of paper to Steve, who read the name, "Dr. William Anixter, Asheville, North Carolina."

"Good luck. I hope he'll be able to help," said the doctor shaking Steve's hand and then Dawn's.

The Hesses thanked Dr. Vittone, then left his office for the quiet, staid beauty of Georgetown.

In his hand, Steve held the missing piece that could make the puzzle come together once and for all.

IV: THEM

Evil? I don't know if they're totally evil. It's like in life. There are some who are evil. But then there are others who are...special. Totally good. And it's those I try to think about especially when things get difficult because I believe it's those that will bring us peace.

Dawn Hess

ONE

Dawn Hess lay in bed with her newborn infant, Amberly Dawn, in her arms. She was weak, but happy, with Steve sitting beside her on the hospital bed.

She smiled weakly.

"I'm so happy the baby is okay."

Dawn gently pulled the pink blanket away, so Steve could see his healthy six pound, eleven ounce baby girl.

"She's a fine looking baby," Steve beamed.

"And I was so scared; *so scared*, Steve, that she wouldn't be normal."

"I know," he comforted. "I know."

"Have you seen my mom and dad?"

He nodded with a proud smirk.

"They're ecstatic. Especially your dad."

"Did you tell them we'd decided to go through with it; the hypnosis, I mean."

"They support it. Think it's the right decision and wanted you to know that."

She looked to him solicitiously.

"And your folks? What do they think?"

He shrugged.

"Diane says she's for getting to the bottom of it if we can, though I know deep down she's terrified of what might happen. And Wolfy," his lips puckered as he struggled for the words, "well, Wolfy still leans toward the military explanation despite what Dr. Vittone said. But one way or the other, he knows things can't go on this way and is for trying to put it to rest, if that's possible."

Dawn's eyes grew wary; unsure as she took Amberly close to her bosom.

"And what about you, Steve? Are you sure this is what's best?"

"I am," he said with conviction. "It's the only way and for better or worse, there's a chance we'll finally know where we stand."

She kissed the baby's cheek gently.

"I'd hoped somehow it would all just go away. But it's not going to stop. They're never going to stop, unless we do something."

Steve reached for her hand.

"I know. But I'm hopeful Dr. Anixter can help. I've done some homework. He's received degrees from George Washington University and UCLA. He's currently Director of Mountain Center of Psychiatry in North Carolina. There's no one better to go to for this kind of thing," he said, determination building like a wave inside him as he spoke, "and if he can just enlighten us; tell us something that could help to make sense of it all..." He sighed heavily. "Well, that's all I'm asking. That's all either of us can ask," he told her.

She considered his words. Then, some other pattern of thought seemed to overshadow what he'd said, rendering it insignificant, even small-minded.

Dawn tugged the bedsheet that partially covered her and Amberly. She carefully unwrapped the soft, cotton receiving blanket that was wound around the baby, so meticulous and tight. Finally, she handed the infant over to him, now clothed only in a diaper, tiny legs pedalling in the air, but uncrying as he took her into his arms.

"This is a 'special' baby, Steve; do you understand that?

She will grow to be a 'special person' who will know things about 'Them' that others don't."

Steve gazed into the tiny blue eyes of his newborn daughter.

Deep in his soul, he knew Dawn was right.

Barstow, California
October 10, 1991
1:05 a.m.

It was two months later while Steve and Dawn were visiting Wolfy and Diane in Barstow that the most salient in a string of occurrences since the birth of Amberly happened.

Steve, Jr. and Bethany, who was now two years old, were sleeping in a room adjoining their grandparents' bedroom. Dawn had put Amberly to bed in their room and was curled up on the couch with Steve watching a late night movie. Dawn put down the Diet Coke she was drinking and looked up to Steve. She felt it and he did, too. A chilling sensation plying its way up the small of her back, until her entire body was left tingling!

Inexplicably, they rose simultaneously, then padded to the sliding glass door that overlooked a fenced backyard and a fathomless view of the nightime sky.

In silence they watched as the stars began moving much as they had in the Mojave that night. And rafts of dark, gray clouds, identical to those that shrouded the huge mothercraft in the desert, began to form in a tumult of roaring motion; rushing its way toward them.

It was at that moment that the baby began crying. It was sudden and violent like they had never heard her cry before.

Wordlessly, Steve and Dawn walked into the darkened bedroom where Amberly lay screaming. Dawn took the infant into her arms, then returned with Steve at her side, through the living room with the TV still squawking, beyond the plateglass door and into the backyard.

The moonless sky was a jumble of motion; hundreds of bright, shiny stars now blinking as if to one another as the churning cauldron of dark clouds stopped them in their tracks.

As if by instinct, Dawn held the baby out into the cool, night air and it stopped. Everything. As if the world was suddenly and totally shut off. Not a sound could be heard. The television stopped playing. The baby stopped crying. The clock in the room stopped ticking. The clouds and stars ceased all movement.

After an unknown period of time, Steve and Dawn returned to their room. They placed Amberly back into her crib, then lay down in bed together.

"It was 'Them'," Dawn said blandly, moments before they fell into a deep and dreamless slumber.

Upland, California
November 20, 1991
6:15 a.m.

Steve pulled his Honda Accord into the driveway of his parents-in-law with Dawn beside him and the three children in the back, both Bethany and Amberly in safety seats.

Five year old Steve, Jr. was complaining.

"I don't want to stay with Bethy and Amberly," he yelped. "I want to go with you!"

"Steven, sweetheart, Mommy and Daddy are going far away on an airplane," she explained, an edge of anxiety creeping into her voice. "Little guys can't go. It's only for grown-ups like us."

"But you never take me anywhere," he whined, "Besides, I'm not little. I'm big."

Steve shook his head good-naturedly.

"Well, whether you're big or little, there are two things I know," he said, turning off the ignition and unbuckling his seatbelt. "One, is that you're staying with your grandmom and grandad. Two, is that we're going to miss our plane if we don't get you dropped off and our tails on the road," he added for Dawn's benefit.

They exited the car from both sides, lifting kids and travel bags loaded with diapers and formula and Ninja Turtle toys.

Bonnie stood at the open door, wide awake, her blond, permed hair already brushed and just-so despite the early hour. Her blue, twinkling eyes seemed nervous and wary as Dawn handed her Amberly with Steven, Jr. trailing behind, looking glum.

Steve, feeling harried and concerned about making the flight, led Bethany toward her grandparents, his arms full of basic necessities for three small children.

Inside the house stood Ed. Sensing the rush, he put his cup of coffee to the side, then extended his long arms toward Steve.

"Here, let me help you with that..."

In a matter of minutes, all three children and all the paraphenalia that went along with them, had been transported from car to home.

"Remember to warm the formula before you give it to Amberly," Dawn reminded. "And don't give Bethany any chocolate no matter how often she asks for it!"

Bonnie nodded, taking the information in one bit at a

time, then repeating it to herself mentally so as to remember.

"Don't worry, dear."

Steve stood beside Dawn, impatiently.

"Dawn, we've got to go..."

At last, Dawn turned to leave, kids behind, as Bonnie and Ed followed them to the door.

"Goodbye," said Dawn to her mom as Steve made his way to the car. "And thanks for taking the children!"

Bonnie just shook her head with tears in her eyes; then Dawn moved forward double-time toward the car.

"Kids," Ed called out as the car doors opened. "I want to tell you, we're proud of what you're doing. It takes a lot of guts. You're very brave."

Steve and Dawn listened, touched by his words.

"And one more thing. Your mom and I have had a lot of time to think about it and we believe you. We believe everything you've told us happened the way you say it did!"

Steve and Dawn waved one final time as they entered the car, their faces flushed with emotion. Steve turned the ignition. The engine started.

They were off to LAX airport and the most significant meeting of their lives.

TWO

Hypnotic Sessions
Mountain Center of Psychiatry
Asheville, North Carolina
November 21, 1991

Laurie, my wife, and I met the Hesses on Thursday morning, November 21 at 7:45 a.m. at the Asheville Marriott. Steve was dressed in jeans, a collared sport shirt and loafers; Dawn wore a pair of black slacks, a long print sweater and flats.

Both were nervous as we waited for a cab that seemed interminably long in getting to the hotel. The wait accentuated everyone's anxiousness about the day—including mine.

At 8:05 a.m. the cab arrived and took us the five mile drive to the Mountain Center of Psychiatry. Laurie sat in the front with the driver. Dawn, Steve and I were in the back.

"So how are you two feeling about this today?" I asked, cautiously.

The word on everyone's mind came from Dawn's lips.

"Nervous," she shivered. "But anxious to get it all over, really."

"Same here," voiced Steve. "We've been living with this thing so long and in so many ways that I've never been more ready to face it head on."

Within a few minutes, the car stood before the Center. I paid the cab driver. The four of us walked up the long stone path leading to the entrance.

Inside, Dr. Anixter greeted us cordially. He was a heavy-set, bearded man in his forties with a quiet, depthful way about him that bespoke his professionalism and experience. After the formalities had been gone through he said simply, "Please come this way."

The doctor's session room was comfortable and old-fashioned with a sofa and soft cushion chairs. The painted walls and pine panelling were done in muted beige and brown colors.

He introduced us to the medical student who would be operating the video equipment, then positioned Laurie and me off to one side and out of his subjects' vision.

Dr. Anixter began by explaining the clinical nature of his approach to the session. His questions would be worded in neutral terms so as to suggest nothing: he would make it clear from the onset to his subjects that an answer such as "I don't know" or "I'm not sure" were acceptable in order to discourage unsure or enhanced responses by his subjects while in a hypnotic state.

The session began.

9:15 a.m.

Dr. Anixter looked from his armchair to Steve Hess, the first to be induced. He asked him to concentrate on a discreet spot on the ceiling, then took him through the hypnotic procedure. It concluded with him walking to Steve, lifting his arm by the wrist, then dropping the limp appendage down onto his lap.

Dr. Anixter then performed the identical procedure on Dawn so that by 10:05 a.m. both subjects were under and totally responsive to his voice and the commands he put forward.

The doctor opened with probes concerning familiar, known events during their experience such as the period when they felt the truck being lifted into the craft, then focused on each of their individual accounts just prior to and including the missing four hours.

"Steve, I'm going to begin with you," he stated in a firm, rapt voice. "Focus on a period of time you previously described as being black. Go back a few minutes before that time."

Everyone in the room watched and waited as Steve Hess began.

"I'm sitting at the end of the bed in the back of the camper looking out the back window."

"What do you remember? Tell me in your own words."

"Things have calmed down. The smaller Beings have left and scattered back through the desert. The activities outside have subsided though I could still see that 'searcher' moving out over the surface of the desert."

"What else? What else do you see?"

His eyelids fluttered furiously as he watched, as if viewing a movie.

"Next thing were Beings. The larger Beings approached the back of the truck. Seems like they were moving toward the back wanting to take a look. They just keep on coming by and I can really smell their thick, musty sulphury odor and feel a tingling sensation in my belly when they were close."

"How are you feeling?" Anixter asked.

"I wanted to reach out and offer my hand and touch them. But Dawn was too scared, so I just sat watching and asking, 'why'? Wanting to jump out of the truck, offer my hand in peace. I'd gotten over feeling scared. *I wanted a physical encounter.*"

"Where did you think you were at this point?"

"Same piece of land, but it seemed we were just not on

earth. Like someone had taken a giant shovel, scooped up a huge plot of earth, then sucked it up toward the center of the ship. Even when we went to sleep we knew it was different. The temperature was colder and it felt like we were floating with the earth we were on, five or six feet thick, then just air underneath." He spoke now with certainty and resentment. "We were inside the craft all right! Feeling like being in a floating museum or a vast, vast collection cavity."

"What's the next thing you remember?"

"I thought it was over for now and time to rest. I said to Dawn, 'It's time to lay down and sleep.' We had a plastic container in back and we both urinated in it because we had to and were too scared to leave the camper. I got back in bed fully clothed with boots on, stretched out crosswise on the bed. Kissed Dawn goodnight, their lights still beaming through the window, then fell asleep."

"And afterward?"

"Waking up all of a sudden in the morning and having sun fill the camper, thinking 'Where am I? What happened?'"

Anixter nodded, expecting the blocking Steve was experiencing. He continued.

"Take a deep breath. Let your mind search. Let the memory come back on its own, then speak out and share it with us. Take your time."

Laurie and I sat on the edge of our seats in anticipation. What would come of this probe into the four hours of missing time? Already, Anixter had established they were inside the ship at the time they fell asleep.

At last, Steve Hess spoke. His face was a mask of concentration. His vision was hazy for the period, though obviously he was struggling to see it.

"I see bright, bright blinking lights...I'm on my back encircled by people looking at me. I see their outlines in the light, strange shadows, but it's so bright..."

"Go on, please," persisted Anixter. "Concentrate. Take a deep breath. Let it out and see if you can remember more..."

"I'm fighting...fighting and resisting them in the camper." Steve is suddenly agitated. "Dawn is grasping for my hand as she's being pulled away...They're white; white all over and I'm fighting." He throws his head to one side. "I want to remember things about that, but the memories don't seem concrete. More like images..."

Suddenly, Dawn's voice rang out with a volume and hysteria that set our teeth on edge.

"Bug eyes!" she screamed.

Anixter swung around to her.

"Say more..."

"It's one of the white ones," she said, perplexed. Then, in a panicked voice, "Can't move. I'm just *scared.*" She began crying. "I want to go home now," she wept, her chest heaving as terror rose up inside her. "Long fingers," she added, suddenly tense as she sees it. "There's three. Around my face. Almost touching, but not touching it."

"What's on the other side of the fingers?" Anixter prodded.

"Light. White, white light. I remember this hallway all lit up in blue lights even in the ground. I guess I'm walking, but it doesn't feel like I'm walking."

"Go on, please."

"It's like a tunnel. Long. No room. Just a long tunnel of blue lights. Until finally, there's an opening. Nothing in there."

"Who are you with?"

"The white guy."

"Where is your husband?"

"I don't know *where* he is," she moans. "I can feel him in my mind, though. I feel like I'm linked to him. We can't talk, but we both know we're going through the same thing."

Anixter sat forward in his chair. "Describe him. The white one."

She concentrates. Her eyes flutter in rapid, sudden motions.

"Big head. Real white. I can't tell what they're made of. Like skin so transluscent it glows. But you can't see veins and things." She strained to see. "I guess he's wearing a shiny covering, too. I don't know what kind of material that is." She hesitated. "I see something on his chest. It's like an arrow. Just the top of an arrow."

"What else?"

"Just three fingers and a thumb. Really shiny. Really long," she elaborated and then with fear and distaste sputtered, "I just hate his eyes. So dark you can't even see them. Can't see the parts like our eyes. They're big, then they go skinny at the ends. That tiny nose. Practically just two holes. Little tiny mouth, but I don't think there are teeth or even lips. A little chin. No hair. Skinny body like a five year old's. Almost my height."

"What happens at this point? What does he say or do, if anything?"

"I'm supposed to just lie on the table. It's okay." She thought for a moment. "I don't know why he keeps looking in my eyes. He puts his fingers on my face just like he did at home that time when he burned me," she explained. "He just moves his hand about five inches from my face, then goes down my body, the length of it."

"Please continue..."

"I'm not supposed to remember anything else. I only remember when they want me to."

"Were you told that?"

"I don't know."

"How do you know when it's all right to remember?"

"I don't know,"

"If you have some way of knowing when it's all right to remember, what does it feel like?"

The words spewed from out of Dawn's mouth like poison. "*Control*!"

"Say more."

"They always know where we are," she spat out angrily. "They always know what we're thinking. They think they can do just anything they want!"

"Please continue. Say what you're comfortable with. What would you like us to know?"

"I don't think they want to hurt us," she spouted. "But I know they can and they will if they have to."

At this point I jotted a handwritten note, hoping the information might help jog Dawn's memory, then handed it to Dr. Anixter.

'You may want to ask about the two marks she found on her neck the morning after the incident.'

Anixter took the note, read it and nodded.

"Did you find two marks on your neck the morning after your experience that hadn't been there before?"

She nodded.

"How did they get there?"

Dawn swallowed hard.

"I was laying on the table," she answered tensely. "And there was this silver wand and it had two tiny openings on the end. They just put it on my neck. It didn't hurt." Then, remembering, "That's what was going over my body, not just his hand. Real shiny."

"What did it feel like?"

"Like a little buzz. A zap. A little shock of current. That's all I'm going to say."

"Why? Is it dangerous to talk?"

"I don't know."

"Is there more you know that you won't share?"

"I can't tell you!" she growled, threateningly.

Doctor Anixter was puzzled at the brick wall he had run into. He gave Dawn instructions to continue thinking about other pieces of the account that were safe to share, then turned his attention back to Steve.

"The last recollections you told me about, Steve, were 'fighting and resisting' the white Beings." He paused to let the words stir recollections from that moment. "I want you to take a deep breath focusing on that moment and tell me what happens next."

Steve took a deep breath. He let it out. He seemed cormfortable; able to remember again.

"That odor. The same as before from the truck. It's everywhere. But now I'm not in the truck. I'm being led down a tunnel, I don't know where, with all white walls. A light guided by a bunch of people on the side pulling me."

"Where's Dawn now?"

"I'm alone. By myself. It's like we've been together going down at first and are now separated with Dawn taken to another area." He bunched up in his chair suddenly. "I remember trying to get away!"

"That's all right. Calm down," Anixter instructed. "What happens next?"

"Through the tunnel. Lots of bright lights. I'm being led by several of the white Beings into an open room. It's an operating room. They put me on a table in the center." He squirmed uncomfortably in his seat as if struggling to get away.

Dr. Anixter spoke in calm, soothing tones.

"It's all right. No one's going to hurt you. What happens now?"

Steve continued, speaking rapidly; fearfully.

"There's one big, round light in the center staring down at me. I'm laying on my back restrained by my arms and legs looking up and seeing the outline of people or beings looking down on top of me." Steve's eyes closed, then shut more tightly. "The light is so bright it's hard to see. But I'm with them. . . fighting. . .there's an examination probe with an instrument!"

"Go on. . ."

"I never talked about it!"

"It's okay to talk. Tell me, what is it you've never talked about?"

"They insert it in my body cavity. I remember fighting. It hurts...pain in my anus and lower stomach." He flails his arms in the air. "I'm telling them to stop."

"Does it stop?" asked Anixter.

"Yeah, it stops." His head lowers. *"I want to go home. I want to see my kids again."*

Doctor Anixter was pensive as he let the past sixty minutes settle in his mind. Then, he heard Dawn murmur.

"Seems like all that bright light would hurt your

eyes."

"What did you say?" he asked urgently.

"Seems like the lights would hurt your eyes, but they don't."

"Are you alone, Dawn?"

"Except for the white Being."

"Then, what?" he prodded.

"Then into the room with the silver table. That same light as in the hallway. And one big, round white light."

"How did you get on the table?"

"It lowers. I sat on it and laid down. No pillow. No blanket and it didn't hurt."

"How did you know what to do?"

"I just knew...I knew. He can tell me with his mind. No words. He puts ideas in my head."

"Is he in communication with you now?"

"He can if he wants."

"Are you being watched now?"

Dawn's head was angled toward the ceiling as before, eyes closed, still directed at a solitary spot. "They know what I'm doing by probes."

The three observers in the room, heeding every syllable of what was being said, gasped collectively. Anixter had done it! Broken through at least one of the barriers that separated us humans from 'them'.

The doctor spoke slowly, methodically now: "Please explain what you mean by 'probes'?"

Dawn reached up with her right hand, eyes still closed, and began pawing at her neck; the right side, near the jugular.

"What does that mean?"

She felt the side of her neck, running her fingers along the flesh.

"I can see you're touching your neck." he told her.
"What does that mean?"

Dawn continued searching that area of her neck
with her fingers as she spoke.

"That's what they shot into my neck that time on
the table with the silver wand."

Anixter paused for a moment as he considered his
next question and how best to position it, then said:

"When you say they're watching you by probes,
what do you mean?"

She spoke with absolute conviction now. The words
shot out like a military command.

"Tracking device put inside of me. That's how they
know where I am and how to communicate."

"How do you know that?"

"I don't know."

"Is it something you're fairly certain of?"

Dawn's hand had still not returned to her side as
she continued, in her hypnotic state, to feel for the
device shot deep into her neck that night.

"Yes. I'm certain," she answered sadly and without
hesitation.

11:10 a.m.

A chill passed through each observer in the room
as Dr. Anixter, Laurie and I attempted to piece the
account together. Without question, they had been on
board the craft in a holding chamber long before they
went to sleep. Apparently drugged or entranced in some
manner, the two were taken from the truck by the white
Beings Dawn had described. A violent struggle ensued
between Steve and their captors as they walked them into

a brightly lit tunnel, then separated them.

Each was then taken into distinct operating rooms where Steve was examined anally with a probing device and Dawn was scanned with a wand-like instrument before a tracking device was implanted deep into the base of her neck.

It all seemed so fantastic. And yet to those familiar with the account and everything the Hesses had been through, what could make more sense? Like animals tracked for study, Steve and Dawn had been trapped, evaluated psychologically and physically, then released as specimans to live in their own natural environment.

Small wonder, these horrible, bone-chilling nightmares of the past two years. Wasn't it obvious why now, psychological explanations aside, they felt watched and monitored. Clearly, they had been.

It was Dr. Anixter's mission now to find out why. The doctor took a deep breath. The session had been arduous from everyone's point of view. An emotional roller coaster that had Dawn, as the doctor was about to begin, quietly sobbing.

"Dawn," Dr. Anixter began once more, "you said before that you don't think the Beings want to hurt you, but they would if they had to. Have they ever hurt either of you?"

"Severe headaches and nightmares," Steve answered swiftly. "Stress at having our emotions pushed to the limit and drained. The procedure was painful ," he went on to explain, "but we feel they've gotten bolder; more aggressive lately."

"How do you mean?"

"Like when they burned Dawn's face," he said resentfully. "They've never done anything like that before!"

"What happened?"

"It was early in the morning," Steve expounded, "maybe 3:00 a.m. when I just lay there staring at the ceiling thinking, 'seems like dark shadows moving on the ceiling, but maybe it's just the trees; then maybe it's not the trees', as I began to feel the presence of the white ones in the room."

"Go on please."

"I opened my eyes and was shocked. I thought my heart had stopped when I saw its face directly over mine. I don't know how he could stand behind me because the headboard was there," he reflected, watching the moment in his mind's eye, "but he did. Then, he ran his long, three fingered hand over Dawn's face, not touching her, then touching her. I saw the welts from his fingers rise up red on Dawn's cheek as I watched and that's when I knew it wasn't a dream and began screaming."

"What else do you remember?" asked Anixter.

"The next morning there were burn marks on her face; a lightning bolt on one cheek and the imprint of three fingers on the other. Dawn called her mom, hysterical, trying to remember how it happened. Then, I remembered the white guy being behind the bed and me snapping pictures with the Polaroid, but it was like it never happened. They always do that," he complained.

"Do you have the pictures?"

"Yes," he answered. "Three of the five came out. But by morning they'd turned black."

Dr. Anixter settled in his chair, attempting now to get them talking more easily about 'safe' topics, so he could move toward the ones they'd been unwilling to discuss earlier.

"Any other experiences where the Beings have injured

or hurt you?"

"Baby Amberly," said Dawn, without thinking. Then, perplexed, "Something to do with Amberly."

"Say more."

"It wasn't the right time in my cycle to get pregnant. I shouldn't have gotten pregnant. But that night Steve looked at me and said, 'You're pregnant' and I said 'I know'. And he said, 'Is it the right time for you to be pregnant?' And I said 'No'." She wracked her brain. "Something to do with the baby."

"What do you mean?"

She sat up straight in her chair, eyes popping open.

"They were there!" she shrieked at the revelations now spilling from her unconscious to her conscious mind. "They were in the room with us that night!"

"Be calm," Anixter coached. "Take a deep breath, then continue."

She did. The instruction seemed to pacify her; sooth her torn nerves and psyche as she settled once again into the cushioned arm chair.

"They were watching us," she recalled, her eyes fluttering in sudden, rapid motions while the images passed before her. "It's like we both knew they were watching, but didn't even care. We had sex and I got pregnant and we both knew it. And we knew that they knew it," she uttered, astonished. "It's like I see them, but I can't see them." Her voice trembled. "I was so scared. I didn't know what Amberly was going to come out being like." Then, suddenly passive. "But she's fine and they left us alone for almost the whole time I was pregnant."

The doctor nodded. The moment was ripe to delve deeper.

"Do either of you have a sense of what it is they want?

Why they've done what they've done?"

It was Steve who answered.

"I keep having these feelings of fear and frustration and contradiction, but I don't think it's over. I don't feel as threatened as I did at first and in a way I believe there may be an underlying source of good."

"How do you mean?"

"Their presence is here and always has been more advanced," he explained. "Their hope is that our world might be able to see or encounter them without mass hysteria through making contact with select people to help raise consciousness and communicate that there are intelligent Beings and a whole existence that parallels our world."

"Where do they come from?"

Steve strained as if trying to recapture the memory of an illusive dream.

"They may have been here all the time," he answered, his eyelids going through a series of rapid motions, "existing spiritually, then becoming physical. They were from other planets, but are more spiritual now. A power and a force."

"And why do they want to make contact?"

"To deliver a message that the world needs to be as one. That if there's war or massive destruction, they will intervene."

"What else do you know?"

He shook his head from side to side, frustrated.

"It's unclear. Too unclear to go on...just thoughts now and images of 'them'."

Dr. Anixter accepted the answer. He turned in his seat toward Dawn, who sat with her body lax, head tilted to one side, lying on her right shoulder.

"Dawn, do you have any idea what these Beings want? Or why they're here?"

It was macabre watching as Dawn sprang up in her chair, suddenly animated and alert. Her voice rather than soft and emotional, took on a new cadence which was clipped and direct. Quite unlike her.

"They want to make contact with the population. Steve and I are specimens; imperfect like the human race. When we're ready to communicate with them face to face, then possibly the world will be, too." Her actions such as edging forward in her chair at the moment seemed mechanical as did her voice as she continued. "They have to study our reactions so they know how to approach us. They don't have emotions like ours, so they need us to teach them. They need to understand humans."

"Do you have a sense as to who they are and where they've come from?" he repeated.

"There are five galaxies. Theirs is the next closest. In order for all five galaxies to work together one day, they have to start and they're starting with us, so we'll be united galaxies."

"What else do you know?" Anixter asked, feeling as if he was in contact with someone other than Dawn.

"I know where the universe ends," she said rattling the words off in staccato fashion like rounds from a machine gun.

"Is that something you can put into words?"

Now, there was no denying: something incredible was happening.

"Our universe ends where theirs begins. Our universe ends when all its matter stops mattering to us and starts mattering to them."

Everyone in the room looked to one another stunned

at what they heard and what they were seeing.

For now, sitting on the edge of the out-of-date cushioned armchair sat Dawn Hess, her body rigid and vibrating with newfound energy.

12:17 p.m.

As was earlier agreed, Dr. Anixter and I then switched places so that I would be able to ask questions.

"Would you mind now if Ron asked some questions of you?"

"No," she answered.

I began, cautiously.

"Dawn, why do you think there are things you don't want to remember?"

"I don't know."

"Is there something preventing you from sharing what you know?"

"I can't tell you."

"If these Beings are in contact with you, tell me what you know about them."

"God created our world in his image, but not theirs," she croaked. "Or any of the others."

"Are there many different kinds of Beings?"

"Yes. Different ones. Different hybrids."

"When do you think this communication will happen? Between the aliens and humans?"

"I don't know."

"In my lifetime?"

"How long will you live?" she retorted.

The hairs on the back of my neck raised.

"In a normal man's lifetime?"

"Our children's," she said at last.

It was then that I had the distinct feeling that I was in a chess match of words. Dawn had begun to take satisfaction in the cleverness; the sharpness of her responses.

I tried a new tack; waiting a moment for the dust to settle, then beginning again in a slower, more calculated manner.

"Are we watched as a population regularly?"

"Yes."

"By many different kinds of Beings..."

"...sent on missions from the One Supreme," she said, finishing my sentence.

"What does that mean?"

"There's one supreme Being that controls all of them. He sends missions here. They're not here of their own accord."

"This is like a president, or..."

"Like a god," she responded with crackling directness.

"What else do you know about the One Supreme?"

"The Beings could care less if they were here or not," she continued in that same, clipped and militant voice. "They're just following orders."

I recouped. 'Where do I take this from here?' I asked myself. The answer I came up with was to retreat back to perhaps the most poignant moment of the session.

"You talked about the universe and matter earlier. You said, 'Our universe ends where theirs begins...'"

"...Our universe ends when its matter stops mattering to us and starts mattering to them."

I reflected briefly on what Steve had said concerning their intervention into our world if war or mass destruction seemed imminent.

"What does that mean?"

"I don't know," she answered simply.

"The 'missions' these aliens are on; are they friendly or unfriendly?"

"They're neutral. They could care less."

"What does the One Supreme want?" I asked pointedly.

"For all the galaxies to live harmoniously together."

"Do we have a separate god from the One Supreme?"

"No."

"What else do you know about this entity? Is it a Being like the others?"

"Read the bible," she quipped.

Again, I stopped to strategize a way of getting at that last elusive piece which Dawn had refused to discuss throughout the session.

"Do they know about our project?"

"Sure."

"Do they care if..." I rephrased it. "Well, it's apparently something they don't want to stop..."

Suddenly, the mask dropped. Her face wrinkled in marked distaste.

"They're into everything; just everything!" she fumed. "Movies like E.T. and Close Encounters; television like Star Trek and even comedies like Alf and Mork and Mindy. It's already in everyone's mind, but so deep they don't even know it. All part of their plan to desensitize people for the contact!"

"Is that what you think," I pressed, "or what they've communicated to you?"

She seemed puzzled and surprisingly vulnerable at that moment. It was a question she, too, had grappled with.

"It pops into my head like I know," she speculated as much to herself as to me or anyone. "It's like someone told me, but no one's ever told me. That's when I think I'm going crazy."

I leaned forward coming very near to her.

"Dawn, the part you don't want to talk about. Is it physical or psychological?"

"I don't know."

"Is there some kind of mental block preventing you from discussing it?"

"I don't know," she repeated, stubbornly.

I stayed on it, equally tenacious.

"The part you don't want to talk about," I asked. "Is it important or unimportant to our project?"

She shook her head in the negative.

"I don't know."

I was getting closer; beginning to feel the resistance weaken.

"Did Steve have a similar experience?"

Her eyes fluttered rapidly; her entire body began to tremble.

"He couldn't have had the very same experience because he's not a woman." She hesitated, then added, "But Laurie knows."

Dr. Anixter seized the moment.

"Are you saying you told Laurie what happened?"

"No, but she knows."

We turned to Laurie who sat to the left of Dawn in the corner of the room, non-plussed.

"They used their probe to examine me inside and out," she spouted, hurt and angry. "They might as well have raped me!" Then, once again evenly, almost as an afterthought, "That's it. That's the part I didn't want to tell you."

Dr. Anixter switched places with me at this point and took charge of the session. Step by step, he brought Steve and Dawn out of their hypnotic state, reassuring them that the information they had discussed was important and needed to be talked about; instructing them that participating in the session was right and good.

The final instruction the doctor gave was that they would not suffer from pangs of guilt or anxiety; that they would come out of their trances refreshed and psychologically disposed toward a feeling of overall well-being.

Individually they awakened, Steve first and then Dawn. Neither had any recollections of what had occurred or been said while in the hypnotic state. Despite the fact that they had been under for nearly four hours, they believed mere moments had passed since the doctor induced them.

Steve stretched his long, muscular body, yawning and obviously stiff from being in one position for this length of time.

Dawn noticed the wetness on her cheeks and wiped the mascara that had smeared from the corners of her eyes. She fluffed her hair, then looked to Dr. Anixter, sincere and ardent.

"So, what do you think, Doctor?" she asked timidly. "Are we...crazy?"

Anixter was equally earnest when he answered, "No, Dawn. Neither one of you is crazy."

2:00 p.m.

Sandwiches and sodas were brought in for the group as Dr. Anixter reviewed and discussed the videotapes with his subjects. Both Steve and Dawn were shocked and pleased by what they heard and saw. At last, they had been vindicated. Their account of the events that transpired on the night of October 21st and morning of the 22nd had been verified under clinical hypnosis. Finally, there was an explanation for the missing four hours and the strange and terrifying events that had followed even to that morning!

I knew what my conclusions were as we sat in Dr. Anixter's session room that afternoon. It was time now for him to share with us his professional opinion.

"Well, Doctor," I said at last. "It's been a long, interesting day. How do you see all of this?"

Anixter ran his left hand along the right side of his beard thoughtfully, then reclined his stout frame back in his armchair.

"I sometimes do work for the FBI using retrogressive hypnotic technique for witnesses that may not remember all that they know about a person or event."

His sharp, discerning eyes scanned the faces before him.

"In those situations, if there is doubt or needs to be speculation on exactly what happened, I lean toward the version that requires the fewest assumptions." He turned the palm of his left hand up in the air, explaining, "In this case, we could say the account is a fabrication." He shrugged. "But why? This situation has endured with them for more than two years. There is no mental illness; so why would two normal, intelligent and adjusted people make up a story that's upset their lives so totally?" He shook his head. "No. For this to be true it requires many far-flung assumptions so that it is unreasonable to believe."

"The other possibility?" I asked, intrigued by the process he was taking us through.

He nodded, placing his hands behind his head, then sank back into the chair, eyes lifting to the ceiling as he collected his thoughts.

"Having eliminated mental illness and the possibility of fabrication, the next would be that the Hesses did, in fact, see and participate in an experience in the Mojave desert that day; perhaps a military exercise, but somehow misinterpretted what happened." His discerning stare dropped now to eye level with each of us. "But, no, the detail here is too exacting and the thoughts and descriptions too tightly organized for that kind of massive misunderstanding." He continued. "And, then, as Doctor Vittone points out, for two people to agree so closely and to have their accounts so divergent from any kind of military exercise I can imagine." He shook his head. "Again, many, many assumptions would be required here."

Dr. Anixter turned to me, then to the Hesses.

"No, Steve and Dawn, I don't think you're crazy. What I think personally and put forward as my professional opinion is that on October 21st and 22nd, 1989, you both had a close encounter of the third kind, unlike any I have ever heard of or seen in my professional experience."

I watched Steve and Dawn's reactions; positive and relieved.

"So, how does that make you feel?" I asked, myself, quietly gratified.

"It makes me feel happy," answered Dawn, "to know after all of this time that it wasn't us and that it really did happen."

Steve smiled. It was reserved, but depthful, rife with satisfaction.

"It makes me feel good," he said, a glint of optimism in his eye that I had never seen before. "Real good about now and about the future."

THREE

La Mirada, California
December 23, 1991
3:30 p.m.

The Christmas tree lights glared red and green and blue as a Lionel train, complete with coal car and caboose, circled and Steve, Jr. and Bethany played, enraptured.

Beyond the ornaments and circle of tracks and children, adults filled the living room of the Hesses' home. Familiar faces, brimming with holiday cheer, talked and drank egg nog and chilled California wine in plastic cups.

There were Wolfy and Diane; Ed and Bonnie; Steve's good friend, Paul Moran; Laurie and me and an assortment of neighbors.

In a far corner of the room, sitting in a sofa chair I spied Dawn with Amberly in her arms. The baby was crying as Dawn attempted to console her. I decided to walk over.

"Nice party," I commented, observing the caring gentleness that Dawn demonstrated with the delicate, fussing infant.

"Yeah," she said, looking up if only for a second. "Steve and I are so glad you could make it."

Soon, Amberly was quiet, resting peacefully on her mom's shoulder.

I took a sip of wine from the clear, plastic cup.

"So, now that it's come this far, how are you and Steve coping?"

"Better," she ventured, "Steve's been offered a new job and we'll probably be moving to Salt Lake City." She chuckled. "Kind of ironic, isn't it?"

"Sure is. Makes you wonder what this whole thing's

been about. Any ideas? I mean, what's your final take on it?"

Dawn smiled contentedly; seemingly at peace with herself and the situation for the first time since I'd known her.

"I don't feel like it's over," she answered, glancing intuitively to Amberly who slept soundly in her arms, "but I don't fear it anymore. It's like that horrible feeling of constant dread has been lifted."

"Why?" I asked, curious. "What's changed?"

In the midst of the conversation and laughter; the loud stories and sounds of children playing, Dawn allowed herself a moment of comtemplation and, perhaps, prayer.

"In many ways, that last session with Dr. Anixter left as many questions as answers. But what I keep remembering is that moment toward the end of our experience when the gremlins were running wild and the illuminated ones were trying to break us and it looked like we were going to be killed. I keep remembering that in the end it was the angel, the Comforter, who dominated."

She looked up to me, her face more radiant; more alive than I had ever seen it. "I think that she's a symbol of something, maybe hope for the future."

The words tumbled in my mind like so many children rolling playfully down green and grassy hillsides.

In the end, it was the angel, the Comforter, who dominated.

EPILOGUE

In the Preface to *Searchers* I suggested that the Hesses' experience in the Mojave desert on October 21st and 22nd, 1989 would, change the way the reader viewed man's place in the world forever. If there has been any shortfall in that expectation after having read their startling and important account, I would like to attribute it to my own inability to convey the magnitude and depth of what happened during that time and in the months to follow.

Like anyone who has read or heard what happened, I was compelled to evaluate each of the alternative explanations based on the facts and my own life experiences. After having rigorously studied their first-hand descriptions along with the tapes of their individual psychiatric sessions, I am personally convinced that an extremely rare and important moment in human development has happened. Rather than some UFO or alien encounter, the Hesses, during those twelve hours and afterward, were allowed to view, conscious and with duration, an entire level of existence that runs contemporary with our own.

Historically, there is undeniable precedent for the existence of this 'other' world. As mentioned in the body of the book, the ancient Indian tribes of the Mojave acknowledged it. But so has the traditional, conventional side of western civilization. The Greeks saw their gods on Mount Olympus manipulating the lives of humans on a giant chess board. Modern, organized religions speak of angels sent down from heaven to counsel prophets and holy men. Philosophers, from Socrates to Camus, struggle with the concepts of free will and predestination. But in the final analysis, these metaphors become incredibly real for perhaps we are, indeed, 'watched' by spiritual mentors who allow us our free will and choice up to a point; that point being self-annihilation.

Described in religious terms, abstracted in philosophical treatise, analogized in the most mundane of pastimes, it all points to the same conclusion. For generations, humans have known subconsciously that there is another level of existence; spiritual "forces" as Steve describes them, locked in a moment to moment struggle over the destiny of humankind. For generations, glimpses into this world, shaded by fear and insecurity, have occurred regularly.

Perhaps soon, we will be in a position to accept the non-traditional, yes, radical viewpoint, so simple and yet so beyond our reach:

There is another world, more advanced and more involved in our future than we can imagine. In truth, they have been here since the beginning of time.

"The devil might be a presence from another universe that wishes to take over our universe. We might be fighting an implaccable enemy out there and the devil might be the agent of that implaccable enemy with God as the tired General fighting that war with his own agents of hope."

Norman Mailer, *David Frost Interview*, 1992

Acknowledgements:

Michael Felber, Sr., Rev. Jeremiah Cullinane,
Eric Shea, Tony Ward, Pete Rodino, Cathy Allgood
and Davey Long; friends and mentors all.

Ron Felber lives in New Jersey with his wife and two children. He was educated at Georgetown University and Loyola University of Chicago. He is the author of four novels, including *The Indian Point Conspiracy* and *The Blue Ice Affair,* and was a recipient of the UPI Award for short stories. Mr. Felber is currently at work on a new non-fiction book set in Turn-of-the-Century Boston.